W9-AMI-096

Aprender ◆ Practicar ◆ Triunfar

Los materiales del estudiante de *Eureka Math*® para *Una historia de unidades*™ (K–5) están disponibles en la trilogía *Aprender, Practicar, Triunfar*. Esta serie apoya la diferenciación y la recuperación y, al mismo tiempo, permite la accesibilidad y la organización de los materiales del estudiante. Los educadores descubrirán que la trilogía *Aprender, Practicar y Triunfar* también ofrece recursos consistentes con la Respuesta a la intervención (RTI, por sus siglas en inglés), las prácticas complementarias y el aprendizaje durante el verano que, por ende, son de mayor efectividad.

Aprender

Aprender de *Eureka Math* constituye un material complementario en clase para el estudiante, a través del cual pueden mostrar su razonamiento, compartir lo que saben y observar cómo adquieren conocimientos día a día. *Aprender* reúne el trabajo en clase—la Puesta en práctica, los Boletos de salida, los Grupos de problemas, las plantillas—en un volumen de fácil consulta y al alcance del usuario.

Practicar

Cada lección de *Eureka Math* comienza con una serie de actividades de fluidez que promueven la energía y el entusiasmo, incluyendo aquellas que se encuentran en *Practicar* de *Eureka Math*. Los estudiantes con fluidez en las operaciones matemáticas pueden dominar más material, con mayor profundidad. En *Practicar*, los estudiantes adquieren competencia en las nuevas capacidades adquiridas y refuerzan el conocimiento previo a modo de preparación para la próxima lección.

En conjunto, *Aprender* y *Practicar* ofrecen todo el material impreso que los estudiantes utilizarán para su formación básica en matemáticas.

Triunfar

Triunfar de *Eureka Math* permite a los estudiantes trabajar individualmente para adquirir el dominio. Estos grupos de problemas complementarios están alineados con la enseñanza en clase, lección por lección, lo que hace que sean una herramienta ideal como tarea o práctica suplementaria. Con cada grupo de problemas se ofrece una Ayuda para la tarea, que consiste en un conjunto de problemas resueltos que muestran, a modo de ejemplo, cómo resolver problemas similares.

Los maestros y los tutores pueden recurrir a los libros de *Triunfar* de grados anteriores como instrumentos acordes con el currículo para solventar las deficiencias en el conocimiento básico. Los estudiantes avanzarán y progresarán con mayor rapidez gracias a la conexión que permiten hacer los modelos ya conocidos con el contenido del grado escolar actual del estudiante.

Estudiantes, familias y educadores:

Gracias por formar parte de la comunidad de *Eureka Math*®, donde celebramos la dicha, el asombro y la emoción que producen las matemáticas.

En las clases de *Eureka Math* se activan nuevos conocimientos a través del diálogo y de experiencias enriquecedoras. A través del libro *Aprender* los estudiantes cuentan con las indicaciones y la sucesión de problemas que necesitan para expresar y consolidar lo que aprendieron en clase.

¿Qué hay dentro del libro Aprender?

Puesta en práctica: la resolución de problemas en situaciones del mundo real es un aspecto cotidiano de *Eureka Math*. Los estudiantes adquieren confianza y perseverancia mientras aplican sus conocimientos en situaciones nuevas y diversas. El currículo promueve el uso del proceso LDE por parte de los estudiantes: Leer el problema, Dibujar para entender el problema y Escribir una ecuación y una solución. Los maestros son facilitadores mientras los estudiantes comparten su trabajo y explican sus estrategias de resolución a sus compañeros/as.

Grupos de problemas: una minuciosa secuencia de los Grupos de problemas ofrece la oportunidad de trabajar en clase en forma independiente, con diversos puntos de acceso para abordar la diferenciación. Los maestros pueden usar el proceso de preparación y personalización para seleccionar los problemas que son «obligatorios» para cada estudiante. Algunos estudiantes resuelven más problemas que otros; lo importante es que todos los estudiantes tengan un período de 10 minutos para practicar inmediatamente lo que han aprendido, con mínimo apoyo de la maestra.

Los estudiantes llevan el Grupo de problemas con ellos al punto culminante de cada lección: la Reflexión. Aquí, los estudiantes reflexionan con sus compañeros/as y el maestro, a través de la articulación y consolidación de lo que observaron, aprendieron y se preguntaron ese día.

Boletos de salida: a través del trabajo en el Boleto de salida diario, los estudiantes le muestran a su maestra lo que saben. Esta manera de verificar lo que entendieron los estudiantes ofrece al maestro, en tiempo real, valiosas pruebas de la eficacia de la enseñanza de ese día, lo cual permite identificar dónde es necesario enfocarse a continuación.

Plantillas: de vez en cuando, la Puesta en práctica, el Grupo de problemas u otra actividad en clase requieren que los estudiantes tengan su propia copia de una imagen, de un modelo reutilizable o de un grupo de datos. Se incluye cada una de estas plantillas en la primera lección que la requiere.

¿Dónde puedo obtener más información sobre los recursos de Eureka Math?

El equipo de Great Minds® ha asumido el compromiso de apoyar a estudiantes, familias y educadores a través de una biblioteca de recursos, en constante expansión, que se encuentra disponible en eureka-math.org. El sitio web también contiene historias exitosas e inspiradoras de la comunidad de *Eureka Math*. Comparte tus ideas y logros con otros usuarios y conviértete en un Campeón de *Eureka Math*.

¡Les deseo un año colmado de momentos "¡ajá!"!

Jill Diniz

Jill Diniz
Directora de matemáticas
Great Minds®

Aprender

Eureka Math®
4.º grado
Módulos 6 y 7

Publicado por Great Minds®.

Copyright © 2019 Great Minds®.

Impreso en los EE. UU.
Este libro puede comprarse en la editorial en eureka-math.org.
2 3 4 5 6 7 8 9 10 CCR 24 23 22

ISBN 978-1-64054-994-4

G4-SPA-M6-M7-L-05.2019

El proceso de Leer-Dibujar-Escribir

El programa de *Eureka Math* apoya a los estudiantes en la resolución de problemas a través de un proceso simple y repetible que presenta la maestra. El proceso Leer-Dibujar-Escribir (LDE) requiere que los estudiantes

1. Lean el problema.

2. Dibujen y rotulen.

3. Escriban una ecuación.

4. Escriban un enunciado (afirmación).

Se procura que los educadores utilicen el andamiaje en el proceso, a través de la incorporación de preguntas tales como

- ¿Qué observas?

- ¿Puedes dibujar algo?

- ¿Qué conclusiones puedes sacar a partir del dibujo?

Cuánto más razonen los estudiantes a través de problemas con este enfoque sistemático y abierto, más interiorizarán el proceso de razonamiento y lo aplicarán instintivamente en el futuro.

Contenido

Módulo 6: Fracciones decimales

Módulo 7: Exploración de medición con multiplicación

© 2019 Great Minds® eureka-math.org

4.° grado

Módulo 6

Nombre _____ Fecha _____

1. Sombrea las primeras 7 unidades del diagrama de cinta. Cuenta en décimas para marcar la recta numérica utilizando una fracción y un decimal para cada punto. Encierra en un círculo el decimal que representa la parte sombreada.

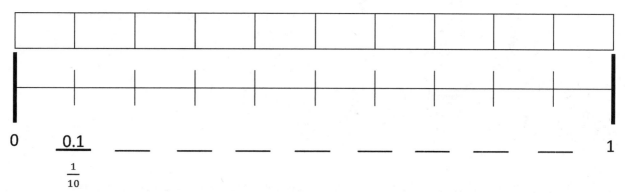

0   0.1   ___   ___   ___   ___   ___   ___   ___   ___   1

$\frac{1}{10}$

2. Escribe la cantidad total de agua en forma de fracción y forma decimal. Sombrea la última botella para mostrar la cantidad correcta.

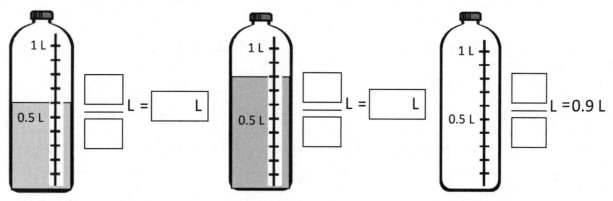

3. Escribe el peso total de la comida en cada báscula en forma de fracción o forma decimal.

| kg | $\frac{8}{10}$ kg | kg |

EUREKA MATH®

Lección 1: Usar la medida métrica para representar la descomposición de un entero en décimas.

© 2019 Great Minds®. eureka-math.org

3

4. Escribe la longitud del insecto en centímetros. (El dibujo no está a escala).

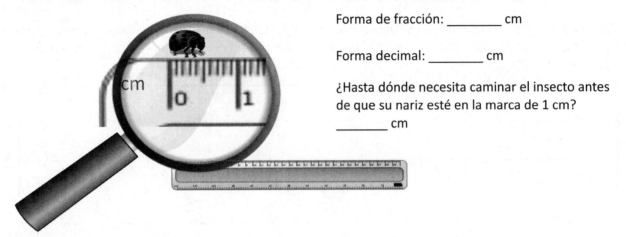

Forma de fracción: _____ cm

Forma decimal: _____ cm

¿Hasta dónde necesita caminar el insecto antes de que su nariz esté en la marca de 1 cm?
_____ cm

5. Llena el espacio en blanco para hacer que el enunciado sea verdadero en forma de fracción y forma decimal.

a. $\frac{8}{10}$ cm + _____ cm = 1 cm   0.8 cm + _____ cm = 1.0 cm

b. $\frac{2}{10}$ cm + _____ cm = 1 cm   0.2 cm + _____ cm = 1.0 cm

c. $\frac{6}{10}$ cm + _____ cm = 1 cm   0.6 cm + _____ cm = 1.0 cm

6. Conecta cada cantidad expresada en forma de unidad a su fracción equivalente y formas decimales.

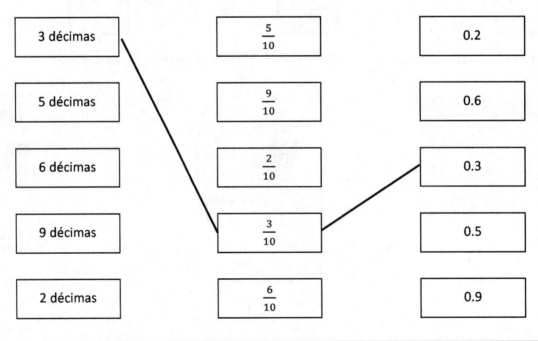

Lección 1:   Usar la medida métrica para representar la descomposición de un entero en décimas.

© 2019 Great Minds®. eureka-math.org

Nombre _____   Fecha _____

1. Llena el espacio en blanco para hacer que el enunciado sea verdadero en forma de fracción y forma decimal.

a. $\frac{9}{10}$ cm + _____ cm = 1 cm   0.9 cm + _____ cm = 1.0 cm

b. $\frac{4}{10}$ cm + _____ cm = 1 cm   0.4 cm + _____ cm = 1.0 cm

2. Conecta cada cantidad expresada en forma de unidad a su forma de fracción y forma decimal.

3 décimas		$\frac{5}{10}$		0.8
8 décimas		$\frac{8}{10}$		0.3
5 décimas		$\frac{3}{10}$		0.5

Lección 1:   Usar la medida métrica para representar la descomposición de un
              entero en décimas.

© 2019 Great Minds®. eureka-math.org

5

Ayer, la planta de bambú de Ben creció 0.5 centírnetros. Hoy creció $\frac{8}{10}$ centímetros. ¿Cuántos centímetros creció la planta de bambú de Ben en 2 días?

_____

_____

_____

_____

Lee   **Dibuja**   **Escribe**

Lección 2:   Usar la medida métrica y modelos de área para representar décimas
                 como fracciones mayores que 1 y números decimales.

© 2019 Great Minds®. eureka-math.org

Nombre _____   Fecha _____

1. Para cada longitud dada a continuación, dibuja un segmento de recta que coincida. Expresa cada medida como un número mixto equivalente.

   a. 2.6 cm

   b. 3.4 cm

   c. 3.7 cm

   d. 4.2 cm

   e. 2.5 cm

2. Escribe lo siguiente como decimales equivalentes. A continuación, representa y renombra el número como se muestra a continuación.

   a. 2 unidades y 6 décimas = _____

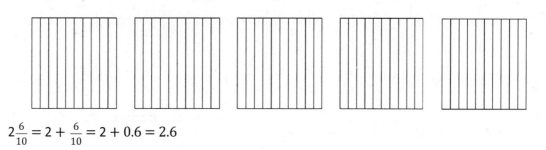

$$2\frac{6}{10} = 2 + \frac{6}{10} = 2 + 0.6 = 2.6$$

Lección 2:   Usar la medida métrica y modelos de área para representar décimas
como fracciones mayores que 1 y números decimales.   **9**

© 2019 Great Minds®. eureka-math.org

b.   4 unidades y 2 décimas = _____

c. $3\frac{4}{10}$ = _____

d. $2\frac{5}{10}$ = _____

¿Cuánto más se necesita para llegar a 5? _____

e. $\frac{37}{10}$ = _____

¿Cuánto más se necesita para llegar a 5? _____

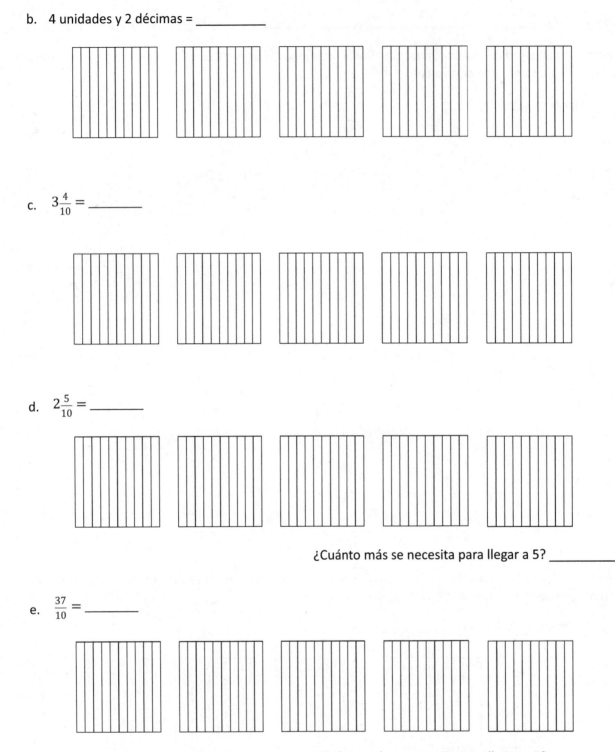

Lección 2:   Usar la medida métrica y modelos de área para representar décimas como fracciones mayores que 1 y números decimales.

© 2019 Great Minds®. eureka-math.org

EUREKA
MATH®

Nombre _____ Fecha _____

1. Para la longitud dada a continuación, dibuja un segmento de recta para que coincida. Expresa la medida como un número mixto equivalente.

   4.8 cm

2. Escribe lo siguiente en forma decimal y como un número mixto. Sombrea el modelo de área para que coincida.

   a. 3 unidades y 7 décimas = _____ = _____

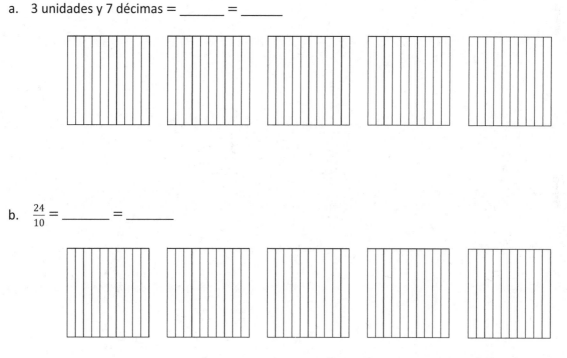

   b. $\frac{24}{10}$ = _____ = _____

   ¿Cuánto más se necesita para llegar a 5? _____

Lección 2:   Usar la medida métrica y modelos de área para representar décimas como fracciones mayores que 1 y números decimales.

11

© 2019 Great Minds®. eureka-math.org

modelo de área de décimas

Lección 2: Usar la medida métrica y modelos de área para representar décimas como fracciones mayores que 1 y números decimales.

13

EUREKA MATH®

© 2019 Great Minds®. eureka-math.org

Ed compró 4 pedazos de salmón con un peso total de 2 kilogramos. Un pedazo pesaba $\frac{4}{10}$ kg y dos de los pedazos pesaban $\frac{5}{10}$ kg cada uno. ¿Cuál era el peso del cuarto pedazo de salmón?

_____

_____

_____

Lee   **Dibuja**   **Escribe**

Lección 3:   Representar números mixtos con decenas, unidades y décimas con discos   15
              de valor posicional, en la recta numérica, y en notación desarrollada.

© 2019 Great Minds®. eureka-math.org

Nombre _____   Fecha _____

1. Encierra en un círculo grupos de décimas para hacer tantas unidades como sea posible.

a. ¿Cuántas décimas en total?	Escribe y dibuja el mismo número usando unidades y décimas.
(0.1)(0.1)(0.1)(0.1)(0.1) (0.1)(0.1)(0.1)(0.1)(0.1) (0.1)(0.1)(0.1)(0.1)(0.1) (0.1)(0.1)(0.1)(0.1)(0.1) (0.1)(0.1)(0.1)(0.1)(0.1) (0.1)(0.1)(0.1) Hay _____ décimas.	▶ Forma decimal: _____ ¿Cuánto más se necesita para llegar a 3? _____
b. ¿Cuántas décimas en total?	Escribe y dibuja el mismo número usando unidades y décimas.
(0.1)(0.1)(0.1)(0.1)(0.1) (0.1)(0.1)(0.1)(0.1)(0.1) (0.1)(0.1)(0.1)(0.1)(0.1) (0.1)(0.1)(0.1)(0.1)(0.1) (0.1)(0.1)(0.1)(0.1)(0.1) (0.1)(0.1)(0.1) (0.1)(0.1)(0.1)(0.1)(0.1) Hay _____ décimas.	▶ Forma decimal: _____ ¿Cuánto más se necesita para llegar a 4? _____

2. Dibuja discos para representar cada número usando decenas, unidades, y décimas. A continuación, muestra la forma desarrollada del número en la forma de fracción y forma decimal como se muestra. El primero ha sido resuelto para ti.

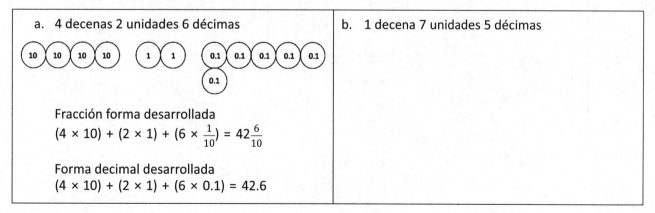

a. 4 decenas 2 unidades 6 décimas	b. 1 decena 7 unidades 5 décimas
(10)(10)(10)(10) (1)(1) (0.1)(0.1)(0.1)(0.1)(0.1) (0.1) Fracción forma desarrollada $(4 \times 10) + (2 \times 1) + (6 \times \frac{1}{10}) = 42\frac{6}{10}$ Forma decimal desarrollada $(4 \times 10) + (2 \times 1) + (6 \times 0.1) = 42.6$	

Lección 3: Representar números mixtos con decenas, unidades y décimas con discos de valor posicional, en la recta numérica, y en notación desarrollada.

17

© 2019 Great Minds®. eureka-math.org

c. 2 decenas 3 unidades 2 décimas	d. 7 decenas 4 unidades 7 décimas

3. Completa la tabla.

Punto	Recta numérica	Forma decimal	Número mixto (unidades y forma de fracción)	Notación desarrollada (forma de fracción o decimal)	¿Cuánto más para llegar a la siguiente unidad?
a.			$3\frac{9}{10}$		0.1
b.	17 18				
c.				$(7 \times 10) + (4 \times 1) + (7 \times \frac{1}{10})$	
d.			$22\frac{2}{10}$		
e.				$(8 \times 10) + (8 \times 0.1)$	

Lección 3:   Representar números mixtos con decenas, unidades y décimas con discos
de valor posicional, en la recta numérica, y en notación desarrollada.

© 2019 Great Minds®. eureka-math.org

Nombre _____   Fecha _____

1. Encierra en un círculo grupos de décimas para hacer tantas unidades como sea posible.

¿Cuántas décimas en total?	Escribe y dibuja el mismo número usando unidades y décimas.
(0.1) (0.1) (0.1) (0.1) (0.1) (0.1) (0.1) (0.1) (0.1) (0.1) (0.1) (0.1) (0.1) (0.1) (0.1) (0.1) (0.1) (0.1) Hay _____ décimas.	 Forma decimal: _____ ¿Cuánto más se necesita para llegar a 2? _____

2. Completa la tabla.

Punto	Recta numérica	Forma decimal	Número Mixto (unidades y forma de fracción)	Forma desarrollada (forma de fracción o decimal)	¿Cuánto más para llegar a la siguiente unidad?
a.	├┼┼┼┼┼┼┼┼┼┤		$12\dfrac{9}{10}$		
b.	├┼┼┼┼┼┼┼┼┼┤	70.7			

Lección 3:   Representar números mixtos con decenas, unidades y décimas con discos
            de valor posicional, en la recta numérica, y en notación desarrollada.

© 2019 Great Minds®. eureka-math.org

19

Punto	Recta numérica	Forma decimal	Número Mixto (unidades y forma de fracción)	Forma desarrollada (forma de fracción o decimal)	¿Cuánto más se necesita para llegar a la siguiente unidad?
a.					
b.					
c.					
d.					

décimas en una recta numérica

Lección 3:   Representar números mixtos con decenas, unidades y décimas con discos de valor posicional, en la recta numérica, y en notación desarrollada.

© 2019 Great Minds®. eureka-math.org

21

Ali está tejiendo una bufanda que tendrá 2 metros de largo. Hasta ahora, ha tejido $1\frac{2}{10}$ metros.

a. ¿Cuántos metros más necesita tejer Ali para completar la bufanda? Escribe la respuesta como fracción y como decimal.

b. ¿Cuántos centímetros más necesita tejer Ali para completar la bufanda?

Lee   Dibuja   Escribe

Lección 4:   Usar metros para representar la descomposición de un entero en centésimas. Representar y contar centésimas.

© 2019 Great Minds®. eureka-math.org

23

Nombre _____ Fecha _____

1. a. ¿Cuál es la longitud en centímetros de la parte sombreada del metro de madera?

1 metro

b. ¿Qué fracción de un metro es 1 centímetro?

c. En forma decimal, expresa la longitud de la parte sombreada del metro de madera.

1 metro

d. En forma decimal, expresa la longitud de la parte sombreada del metro de madera.

e. ¿Qué fracción de un metro son 10 centímetros?

2. Llena los espacios en blanco.

a. 1 décima = _____ centésimas

b. $\frac{1}{10}$ m = $\frac{}{100}$ m

c. $\frac{2}{10}$ m = $\frac{20}{}$ m

3. Utiliza la representación para sumar las partes sombreadas como se muestra. Escribe un vínculo numérico con el total escrito en forma decimal y las partes escritas como fracciones. El primer ejercicio ya está resuelto.

a.

1 metro

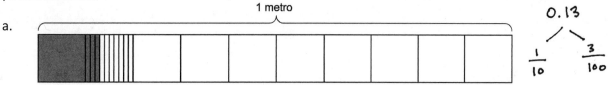

$$\frac{1}{10}m + \frac{3}{100}m = \frac{13}{100}m = 0.13\,m$$

Lección 4: Usar metros para representar la descomposición de un entero en centésimas. Representar y contar centésimas.

25

© 2019 Great Minds®. eureka-math.org

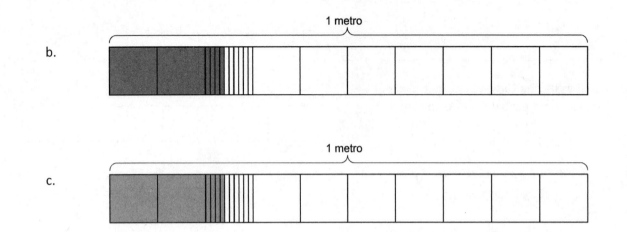

b.

c.

4. En cada metro de madera, sombrea la cantidad indicada. Luego, escribe el equivalente decimal.

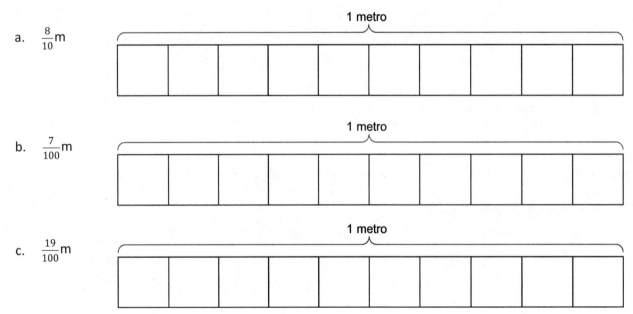

a. $\frac{8}{10}$ m

b. $\frac{7}{100}$ m

c. $\frac{19}{100}$ m

5. Dibuja un vínculo numérico, sacando las décimas de las centésimas como en el problema 3. Escribe el total como el equivalente decimal.

a. $\frac{19}{100}$ m

b. $\frac{28}{100}$ m

c. $\frac{77}{100}$

d. $\frac{94}{100}$

Lección 4:   Usar metros para representar la descomposición de un entero en centésimas. Representar y contar centésimas.

© 2019 Great Minds®. eureka-math.org

Nombre _____   Fecha _____

1. Sombrea la cantidad indicada. Luego, escribe el equivalente decimal.

$\frac{6}{10}$m

1 metro

2. Dibuja un vínculo numérico, sacando las décimas de las centésimas. Escribe el total como el equivalente decimal.

    a. $\frac{62}{100}$m

    b. $\frac{27}{100}$

Lección 4:   Usar metros para representar la descomposición de un entero en
              centésimas. Representar y contar centésimas.

© 2019 Great Minds®. eureka-math.org

27

1 metro

1 metro

1 metro

1 metro

1 metro

diagrama de cinta en décimas

Lección 4:   Usar metros para representar la descomposición de un entero en
              centésimas. Representar y contar centésimas.

© 2019 Great Minds®. eureka-math.org

29

El perímetro de un cuadrado mide 0.48 m. ¿Cuánto mide cada longitud lateral en centímetros?

_____

_____

_____

_____

Lee   **Dibuja**   **Escribe**

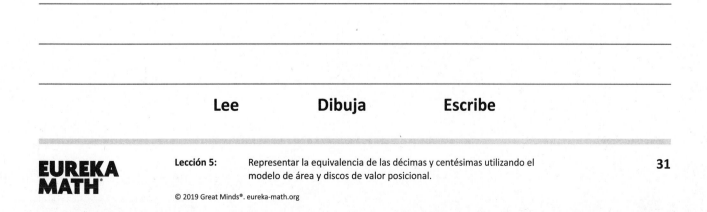

Lección 5:   Representar la equivalencia de las décimas y centésimas utilizando el modelo de área y discos de valor posicional.

31

© 2019 Great Minds®. eureka-math.org

Nombre _____ Fecha _____

1. Encuentra la fracción equivalente usando la multiplicación o la división. Sombrea los modelos de área para mostrar la equivalencia. Registra como decimal.

a. $\dfrac{3\times}{10\times}=\dfrac{}{100}$

b. $\dfrac{50\div}{100\div}=\dfrac{}{10}$

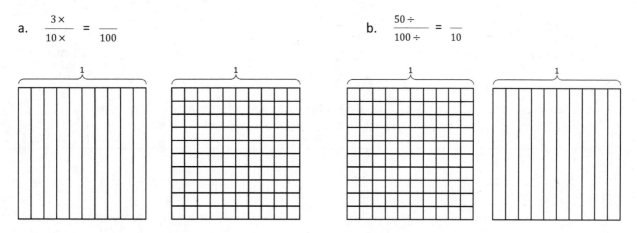

2. Completa los enunciados numéricos. Sombrea la cantidad equivalente en el modelo de área, trazando líneas horizontales para hacer centésimas.

a. 37 centésimas = _____ décimas + _____ centésimas

   Forma de fracción: _____

   Forma decimal: _____

b. 75 centésimas = _____ décimas + _____ centésimas

   Forma de fracción: _____

   Forma decimal: _____

3. Encierra en un círculo las centésimas para componer tantas décimas como sea posible. Completa los enunciados numéricos. Representa cada uno con un vínculo numérico como se muestra.

a.

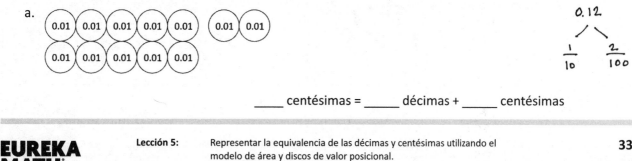

_____ centésimas = _____ décimas + _____ centésimas

Lección 5:   Representar la equivalencia de las décimas y centésimas utilizando el modelo de área y discos de valor posicional.

33

EUREKA MATH®

© 2019 Great Minds®. eureka-math.org

b.

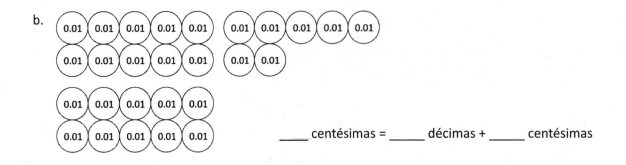

_____ centésimas = _____ décimas + _____ centésimas

4. Usa discos de valor posicional de décimas y de centésimas para representar cada número. Escribe el número equivalente en decimal, fracción y forma de unidad.

a. $\frac{3}{100}$ = 0. _____ _____ centésimas	b. $\frac{15}{100}$ = 0. _____ _____ décima _____ centésimas
c. —— = 0.72 _____ centésimas	d. —— = 0.80 _____ décimas
e. —— = 0. _____ 7 décimas 2 centésimas.	f. —— = 0. _____ 80 centésimas

Lección 5:   Representar la equivalencia de las décimas y centésimas utilizando el modelo de área y discos de valor posicional.

© 2019 Great Minds®. eureka-math.org

Nombre _____ Fecha _____

Usa ambos discos de valor posicional de décimas y centésimas para representar cada fracción.
Escribe el equivalente decimal, y llena los espacios en blanco para representar cada una en forma de unidad.

1. $\frac{7}{100}$ = 0. ____

_____ centésimas

2. $\frac{34}{100}$ = 0. ____

_____ décimas _____ centésimas

Lección 5: Representar la equivalencia de las décimas y centésimas utilizando el modelo de área y discos de valor posicional.

35

© 2019 Great Minds®. eureka-math.org

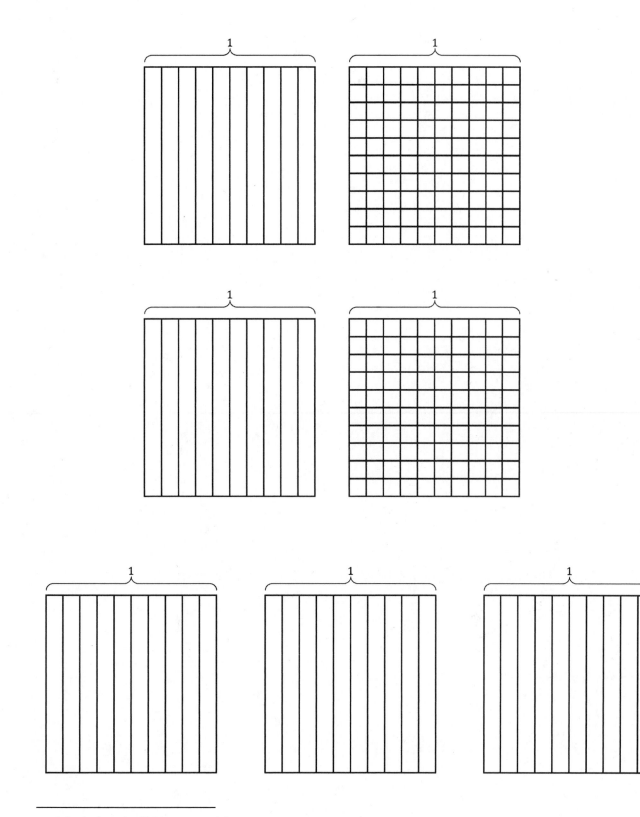

modelo de área de décimas y centésimas

Lección 5: Represesntar la equivalencia de las décimas y centésimas utilizando el modelo de área y discos de valor posicional.

37

© 2019 Great Minds®. eureka-math.org

La tabla muestra el perímetro de cuatro rectángulos.

a. ¿Qué rectangulo tiene el menor perímetro?

Rectángulo	Perímetro
A	54 cm
B	$\frac{69}{100}$ m
C	54 m
D	0.8 m

b. ¿Cuántos metros menos de un kilómetro tiene el perímetro del rectángulo C?

Lee　　　　**Dibuja**　　　　**Escribe**

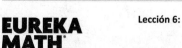

c. Compara los perímetros de los rectángulos B y D. ¿Qué rectángulo tiene el mayor perímetro?
¿Cuánto más grande?

Lee   **Dibuja**   **Escribe**

Lección 6:   Usar el modelo de área y la recta numérica para representar números
mixtos con unidades, décimas, centésimas, en forma de fracciones y
decimales.

© 2019 Great Minds®. eureka-math.org

Nombre _____ Fecha _____

1. Sombrea los modelos de área para representar el número, trazando líneas horizontales para hacer centésimas según sea necesario. Busca el punto correspondiente en la recta numérica. Marca con un punto y registra el número mixto como un decimal.

   a. $1\frac{15}{100}$ = ___ . _____

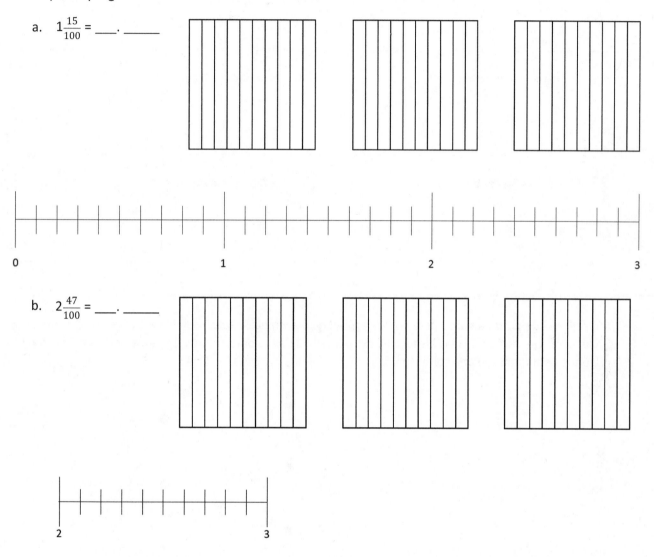

   b. $2\frac{47}{100}$ = ___ . _____

2. Calcula aproximadamente la localización de los puntos en las rectas numéricas.

   a. $2\frac{95}{100}$ b. $7\frac{52}{100}$

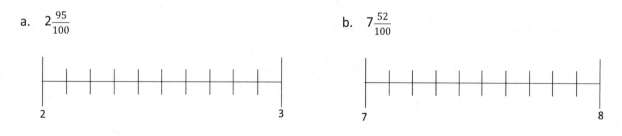

Lección 6:   Usar el modelo de área y la recta numérica para representar números mixtos con unidades, décimas, centésimas, en forma de fracciones y decimales.

© 2019 Great Minds®. eureka-math.org

41

3. Escribe la fracción y decimal equivalente para eada uno de los siguientes números.

a. 1 unidad 2 centésimas	b. 1 unidad 17 centésimas
c. 2 unidades 8 centésimas	d. 2 unidades 27 centésimas
e. 4 unidades 58 centésimas	f. 7 unidades 70 centésimas

4. Dibuja líneas de punto a punto para que la forma decimal coincida tanto con la forma de unidad como con la forma de fracción. Todas las formas de unidad y de fracciones tienen al menos una coincidencia, y algunos tienen más de una coincidencia.

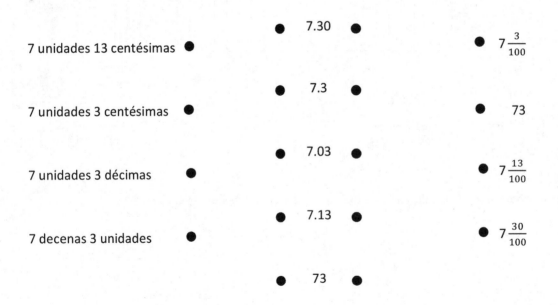

7 unidades 13 centésimas

7 unidades 3 centésimas

7 unidades 3 décimas

7 decenas 3 unidades

7.30

7.3

7.03

7.13

73

$7\frac{3}{100}$

73

$7\frac{13}{100}$

$7\frac{30}{100}$

Lección 6:   Usar el modelo de área y la recta numérica para representar números mixtos con unidades, décimas, centésimas, en forma de fracciones y decimales.

© 2019 Great Minds®. eureka-math.org

Nombre _____   Fecha _____

1. Calcula aproximadamente la localización de los puntos en las rectas numéricas. Marca el punto y etiquétalo como un decimal.

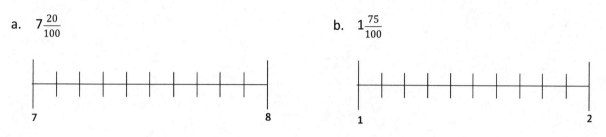

a. $7\frac{20}{100}$ b. $1\frac{75}{100}$

2. Escribe la fracción y decimal equivalente para cada número.

     a.   8 unidades 24 centésimas   b.   2 unidades 6 centésimas

Lección 6:   Usar el modelo de área y la recta numérica para representar números mixtos con unidades, décimas, centésimas, en forma de fracciones y decimales.

© 2019 Great Minds®. eureka-math.org

43

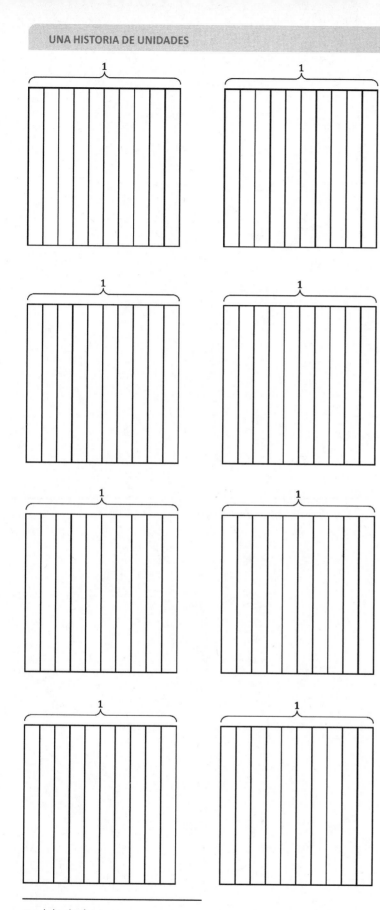

modelo de área

Lección 6:   Usar el modelo de área y la recta numérica para representar números
              mixtos con unidades, décimas, centésimas, en forma de fracciones y
              decimales.

© 2019 Great Minds®. eureka-math.org

45

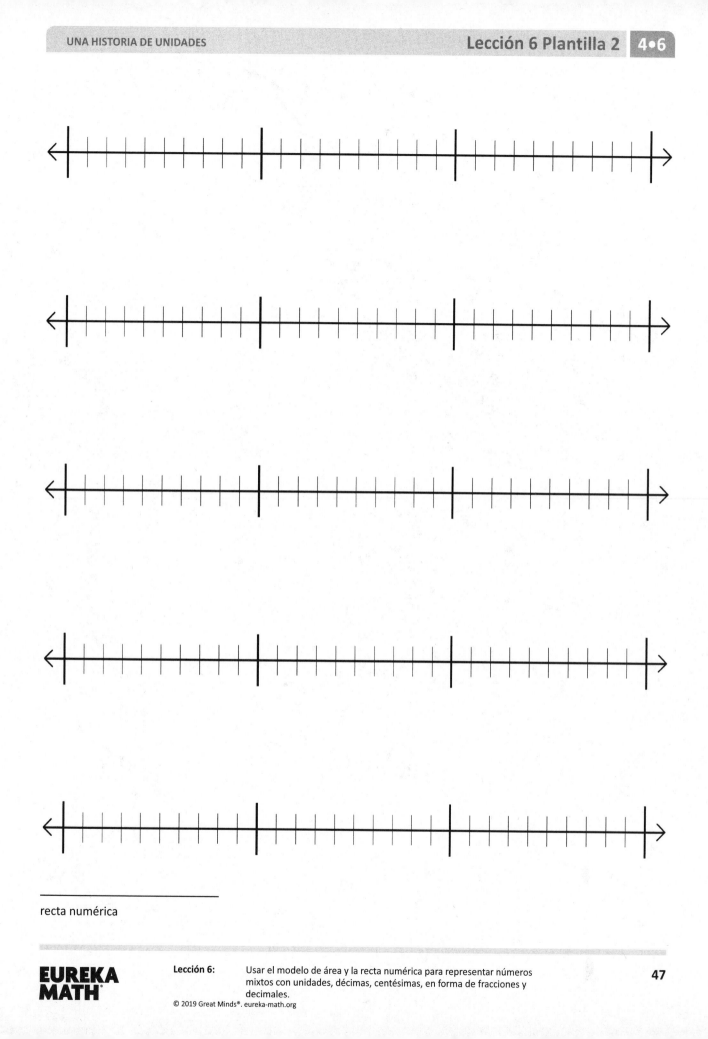

recta numérica

Lección 6:   Usar el modelo de área y la recta numérica para representar números
mixtos con unidades, décimas, centésimas, en forma de fracciones y
decimales.
© 2019 Great Minds®. eureka-math.org

47

EUREKA
MATH®

Utiliza bloques de patrón para crear por lo menos 1 figura con al menos 1 línea de simetría. Dibuja la figura a continuación.

Lee   Dibuja   Escribe

Lección 7: Representar números mixtos con centenas, decenas, unidades, décimas y centésimas, en notación desarrollada y en la tabla de valor posicional.

© 2019 Great Minds®. eureka-math.org

49

EUREKA MATH®

Nombre _____ Fecha _____

1. Escribe un enunciado numérico decimal para identificar el valor total de los discos de valor posicional.

a.

(10) (10) (0.1) (0.1) (0.1) (0.1) (0.1) (0.01) (0.01) (0.01)

2 decenas   5 décimas   3 centésimas

_____ + _____ + _____ = _____

b.

(100) (100) (100) (100) (100) (0.01) (0.01) (0.01) (0.01)

5 centenas   4 centésimas

_____ + _____ = _____

2. Usa la tabla de valor posicional para contestar las siguientes preguntas. Expresa el valor del dígito en la forma de unidad.

centenas	decenas	unidades	·	décimas	centésimas
4	1	6		8	3

a. El dígito _____ está en el lugar de las centenas. Tiene un valor de _____.

b. El dígito _____ está en el lugar de las decenas. Tiene un valor de _____.

c. El dígito _____ está en el lugar de las décimas. Tiene un valor de _____.

d. El dígito _____ está en el lugar de los centésimas. Tiene un valor de _____.

centenas	decenas	unidades	·	décimas	centésimas
5	3	2		1	6

e. El dígito _____ está en el lugar de las centenas. Tiene un valor de _____.

f. El dígito _____ está en el lugar de las decenas. Tiene un valor de _____.

g. El dígito _____ está en el lugar de las décimas. Tiene un valor de _____.

h. El dígito _____ está en el lugar de los centésimas. Tiene un valor de _____.

Lección 7: Representar números mixtos con centenas, decenas, unidades, décimas y centésimas, en notación desarrollada y en la tabla de valor posicional.

© 2019 Great Minds®. eureka-math.org

51

3. Escribe cada decimal como una fracción equivalente. Luego, escribe cada número en forma desarrollada, utilizando la notación decimal y de fracción. El primer ejercicio ya está resuelto.

Forma decimal y fracción	Forma desarrollada	
	Notación de fracción	Notación decimal
$15.43 = 15\frac{43}{100}$	$(1 \times 10) + (5 \times 1) + (4 \times \frac{1}{10}) + (3 \times \frac{1}{100})$ $10 + 5 + \frac{4}{10} + \frac{3}{100}$	$(1 \times 10) + (5 \times 1) + (4 \times 0.1) + (3 \times 0.01)$ $10 + 5 + 0.4 + 0.03$
$21.4 = $ _____		
$38.09 = $ _____		
$50.2 = $ _____		
$301.07 = $ _____		
$620.80 = $ _____		
$800.08 = $ _____		

Lección 7:   Representar números mixtos con centenas, decenas, unidades, décimas y centésimas, en notación desarrollada y en la tabla de valor posicional.

© 2019 Great Minds®. eureka-math.org

EUREKA MATH®

Nombre _____   Fecha _____

1. Usa la tabla de valor posicional para contestar las siguientes preguntas. Expresa el valor del dígito en la forma de unidad.

centenas	decenas	unidades	•	décimas	centésimas
8	2	7		6	4

a. El dígito_____está en el lugar de las centenas. Tiene un valor de _____.

b. El dígito_____está en el lugar de las decenas. Tiene un valor de _____.

c. El dígito_____está en el lugar de las décimas. Tiene un valor de _____.

d. El dígito_____está en el lugar de los centésimas. Tiene un valor de _____.

2. Completa la siguiente tabla.

| Fracción | Forma desarrollada | | Decimal |
	Notación de fracción	Notación decimal	
$422\frac{8}{100}$			
	$(3\times100)+(9\times\frac{1}{10})+(2\times\frac{1}{100})$		

Lección 7: Representar números mixtos con centenas, decenas, unidades, décimas y centésimas, en notación desarrollada y en la tabla de valor posicional.

© 2019 Great Minds®. eureka-math.org

53

centenas	decenas	unidades	.	décimas	centésimas

tabla de valor posicional

Lección 7:   Representar números mixtos con centenas, decenas, unidades, décimas y centésimas, en notación desarrollada y en la tabla de valor posicional.

© 2019 Great Minds®. eureka-math.org

55

Jashawn tenía 5 billetes de cien dólares y 6 billetes de diez dólares en su cartera. Alva tenía 58 billetes de diez dólares bajo el colchón. James tenía 556 billetes de un dólar en su alcancía. Ellos deciden combinar su dinero para comprar una computadora. Expresa la cantidad total de dinero que han utilizado en los siguientes billetes:

a. Billetes de cien, de diez y de uno.

b. Billetes de diez y de uno.

Lee   **Dibuja**   **Escribe**

Lección 8:   Usar la comprensión de la fracción equivalente para investigar los números decimales en la tabla de valor posicional expresados en diferentes unidades.

© 2019 Great Minds®. eureka-math.org

c. Billetes de uno.

Lee   Dibuja   Escribe

Lección 8:   Usar la comprensión de la fracción equivalente para investigar los números decimales en la tabla de valor posicional expresados en diferentes unidades.

© 2019 Great Minds®. eureka-math.org

Nombre _____   Fecha _____

1. Utiliza el modelo de área para representar $\frac{250}{100}$. Completa el enunciado numérico.

   a. $\frac{250}{100}$ = _____ décimas = _____ unidades _____ décimas = __.____

   b. En el espacio de abajo, explica cómo determinaste tu respuesta a la parte (a).

2. Dibuja los discos de valor posicional para representar las siguientes descomposiciones:

   2 unidades = _____ décimas

unidades	·	décimas	centésimas

   2 décimas = _____ centésimas

unidades	·	décimas	centésimas

   1 unidad 3 décimas = _____ décimas

unidades	·	décimas	centésimas

   2 décimas 3 centésimas = _____ centésimas

unidades	·	décimas	centésimas

Lección 8:   Usar la comprensión de la fracción equivalente para investigar los números decimales en la tabla de valor posicional expresados en diferentes unidades.

© 2019 Great Minds®. eureka-math.org

59

3. Descompón las unidades para representar cada número como décimas.

   a. 1 = _____ décimas

   b. 2 = _____ décimas

   c. 1.7 = _____ décimas

   d. 2.9 = _____ décimas

   e. 10.7 = _____ décimas

   f. 20.9 = _____ décimas

4. Descompón las unidades para representar cada número como centésimas.

   a. 1 = _____ centésimas

   b. 2 = _____ centésimas

   c. 1.7 = _____ centésimas

   d. 2.9 = _____ centésimas

   e. 10.7 = _____ centésimas

   f. 20.9 = _____ centésimas

5. Completa la tabla. El primer ejercicio ya está resuelto.

Decimal	Número mixto	Décimas	Centésimas
2.1	$2\frac{1}{10}$	21 décimas $\frac{21}{10}$	210 centésimas $\frac{210}{100}$
4.2			
8.4			
10.2			
75.5			

Lección 8:   Usar la comprensión de la fracción equivalente para investigar los números decimales en la tabla de valor posicional expresados en diferentes unidades.

© 2019 Great Minds®. eureka-math.org

Nombre _____ Fecha _____

1. a. Dibuja los discos de valor posicional para representar la siguiente descomposición:

3 unidades 2 décimas = _____ décimas

unidades	•	décimas	centésimas

b. 3 unidades 2 décimas =_____centésimas

2. Descompón las unidades.

a. 2.6 =_____décimas

b. 6.1 =_____centésimas

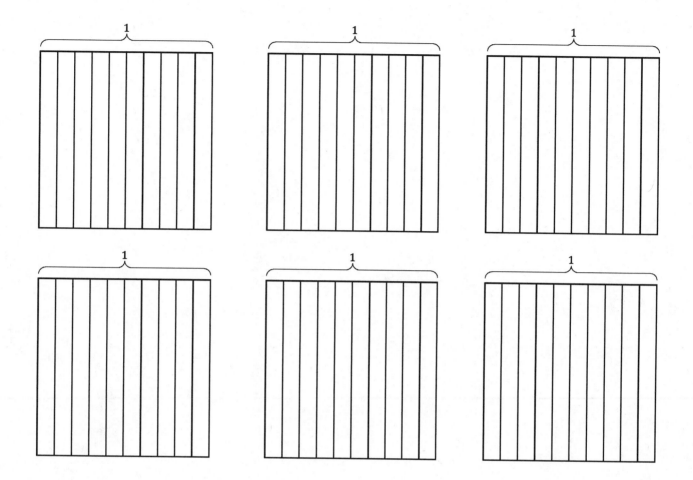

Decenas	Unidades	.	Décimas	Centésimas

modelo de área y tabla de valor posicional

EUREKA MATH®

Lección 8: Usar la comprensión de la fracción equivalente para investigar los números decimales en la tabla de valor posicional expresados en diferentes unidades.

© 2019 Great Minds®. eureka-math.org

63

El perro de Kelly pesa 14 kilogramos 24 gramos. El perro de Mary pesa 14 kilogramos 205 gramos.
El perro de Hae Jung pesa 4,720 gramos.

    a. Ordena el peso de los perros en gramos de menor a mayor.

    b. ¿Cuánto más pesa el perro más pesado que el perro más liviano?

Lee   Dibuja   Escribe

          Lección 9:   Usar la tabla de valor posicional y medidas métricas para comparar decimales   65
                                        y responder a las preguntas de comparación.

                          © 2019 Great Minds®. eureka-math.org

Nombre _____   Fecha _____

1. Expresa las longitudes de las partes sombreadas en forma decimal. Escribe un enunciado que compare las dos longitudes. Utiliza la expresión *más corto que* o *más largo que* en tu enunciado.

a.

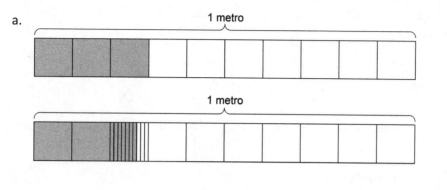

1 metro

1 metro

b.

1 metro

1 metro

c. Enlista las cuatro longitudes de menor a mayor.

2. a. Examina la masa de cada artículo como se muestra a continuación en las escalas de 1 kilogramo. Marca con una X los artículos que son más pesados que el aguacate.

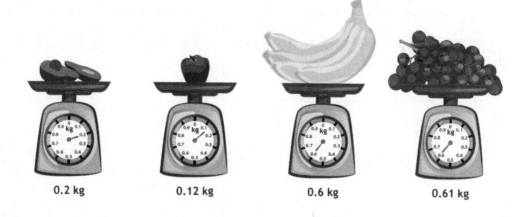

0.2 kg   0.12 kg   0.6 kg   0.61 kg

Lección 9:   Usar la tabla de valor posicional y medidas métricas para comparar decimales y responder a las preguntas de comparación.

67

© 2019 Great Minds®. eureka-math.org

b. Expresa la masa de cada artículo en la tabla de valor posicional.

Masa de la fruta (Kilogramos)

Fruta	unidades	·	décimas	centésimas
aguacate				
manzana				
bananas				
uvas				

c. Completa los siguientes enunciados, usando las palabras *más pesado que* o *más ligero que* en tus enunciados.

El aguacate es _____ que la manzana.

El racimo de plátanos es _____ que el racimo de uvas.

3. Registra el volumen de agua en cada cilindro graduado en la tabla de valor posicional a continuación.

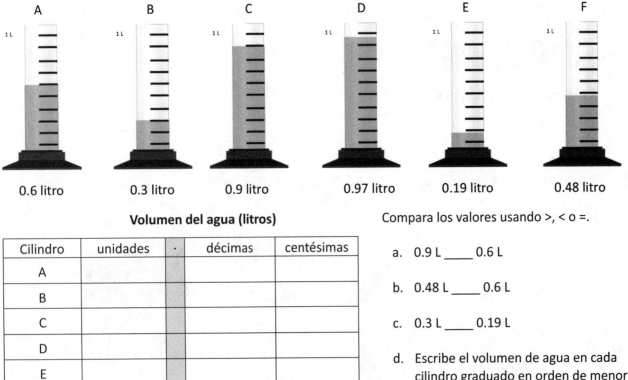

A	B	C	D	E	F
0.6 litro	0.3 litro	0.9 litro	0.97 litro	0.19 litro	0.48 litro

Volumen del agua (litros)

Cilindro	unidades	·	décimas	centésimas
A				
B				
C				
D				
E				
F				

Compara los valores usando >, < o =.

a. 0.9 L _____ 0.6 L

b. 0.48 L _____ 0.6 L

c. 0.3 L _____ 0.19 L

d. Escribe el volumen de agua en cada cilindro graduado en orden de menor a mayor.

© 2019 Great Minds®. eureka-math.org

Nombre _____   Fecha _____

1. a. Doug mide las longitudes de tres cuerdas y sombrea los diagramas de cinta para representar la longitud de cada cuerda como se muestra a continuación. Expresa, en forma decimal, la longitud de cada cuerda.

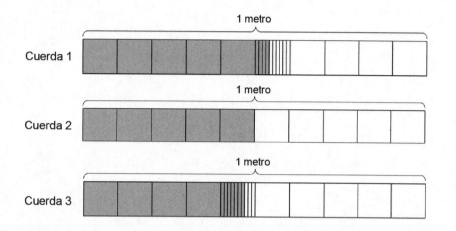

    b. Enlista las longitudes de las cuerdas en orden de mayor a menor.

2. Compara los valores debajo usando >, < o =.

    a. 0.8 kg _____ 0.6 kg

    b. 0.36 kg _____ 0.5 kg

    c. 0.4 kg _____ 0.47 kg

Lección 9:   Usar la tabla de valor posicional y medidas métricas para comparar decimales
             y responder a las preguntas de comparación.

© 2019 Great Minds®. eureka-math.org

69

Masa de las bolsas de arroz (kilogramos)

Bolsa de arroz	unidades	.	décimas	centésimas
A				
B				
C				
D				

Volumen de líquido (litros)

Cilindro	unidades	.	décimas	centésimas
A				
B				
C				
D				

registro de medida

Lección 9: Usar la tabla de valor posicional y medidas métricas para comparar decimales y responder a las preguntas de comparación.

71

© 2019 Great Minds®. eureka-math.org

En la clase de ciencias, el matraz de 1 litro de Emily contiene 0.3 litros de agua. El matraz de Ali contiene 0.8 litros de agua y el matraz de Katie contiene 0.6 litros de agua. ¿Quién puede verter toda su agua en el matraz de Emily, sin pasarse de 1 litro, Ali o Katie?

_____

_____

_____

_____

Lee   **Dibuja**   **Escribe**

Lección 10:   Usar modelos de área y la recta numérica para comparar números decimales, y registrar las comparaciones usando <, > y =.

73

© 2019 Great Minds®. eureka-math.org

Nombre _____   Fecha _____

1. Sombrea los modelos de área a continuación, descomponiendo décimas según sea necesario, para representar los pares de números decimales. Llena el espacio en blanco con <, > o = para comparar los números decimales.

a. 0.23 _____ 0.4

b. 0.6 _____ 0.38

c. 0.09 _____ 0.9

d. 0.70 _____ 0.7

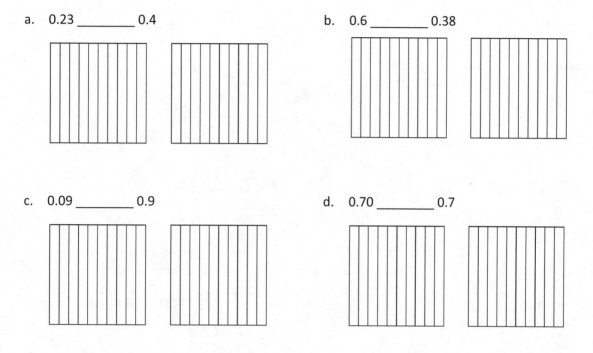

2. Localiza y marca los puntos para cada uno de los números decimales en la recta numérica. Llena el espacio en blanco con <, > o = para comparar los números decimales.

a. 10.03 _____ 10.3

b. 12.68 _____ 12.8

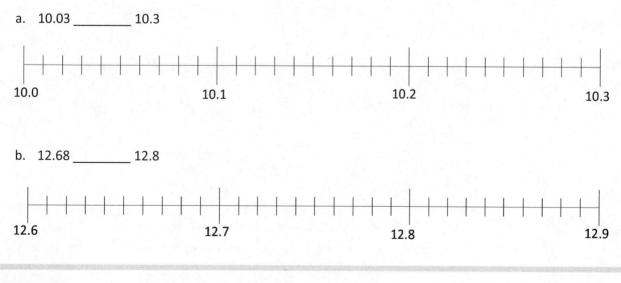

Lección 10: Usar modelos de área y la recta numérica para comparar números decimales, y registrar las comparaciones usando <, > y =.

75

© 2019 Great Minds®. eureka-math.org

3. Usa los símbolos <, > o = para comparar.

a. 3.42 _____ 3.75

b. 4.21 _____ 4.12

c. 2.15 _____ 3.15

d. 4.04 _____ 6.02

e. 12.7 _____ 12.70

f. 1.9 _____ 1.21

4. Usa los símbolos <, > o = para comparar. Utiliza imágenes cuando sea necesario para resolver.

a. 23 décimas _____ 2.3

b. 1.04 _____ 1 unidad y 4 décimas

c. 6.07 _____ $6\frac{7}{10}$

d. 0.45 _____ $\frac{45}{10}$

e. $\frac{127}{100}$ _____ 1.72

f. 6 décimas _____ 66 centésimas

Lección 10: Usar modelos de área y la recta numérica para comparar números decimales, y registrar las comparaciones usando <, > y =.

© 2019 Great Minds®. eureka-math.org

Nombre _____   Fecha _____

1. Ryan dice que 0.6 es menor que 0.60, ya que tiene menos dígitos. Jessie dice que 0.6 es mayor que 0.60. ¿Quién está en lo correcto? ¿Por qué? Utiliza los modelos de área de abajo para ayudar a explicar tu resnuesta.

0.6 _____ 0.60

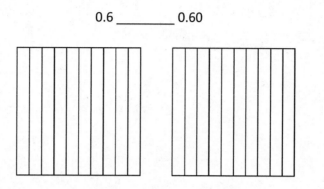

2. Usa los símbolos <, > o = para comparar.

   a. 3.9 _____ 3.09

   b. 2.4 _____ 2 unidades y 4 centésimas

   c. 7.84 _____ 78 décimas y 4 centésimas

    Lección 10:   Usar modelos de área y la recta numérica para comparar números decimales,   77
y registrar las comparaciones usando <, > y =.

© 2019 Great Minds®. eureka-math.org

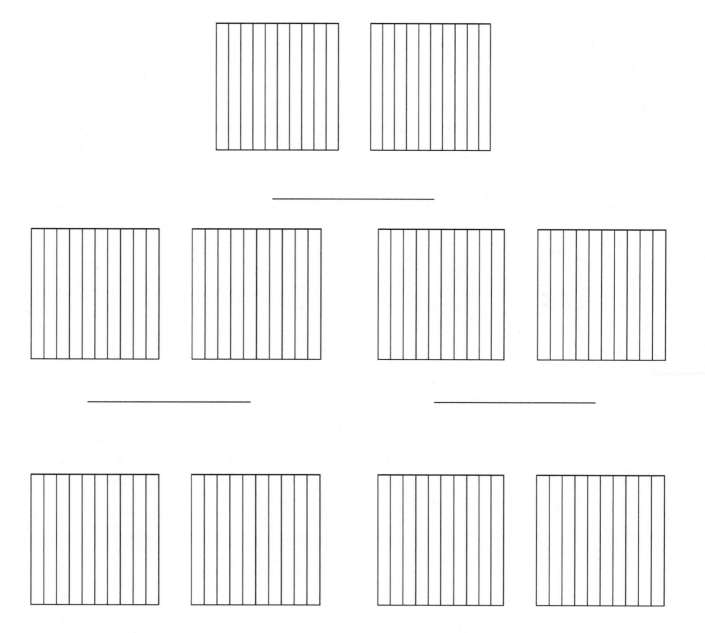

comparación con modelos de área

Lección 10: Usar modelos de área y la recta numérica para comparar números decimales, y registrar las comparaciones usando <, > y =.

79

© 2019 Great Minds®. eureka-math.org

Durante la costura, Kikanza cortó 3 tiras de tela de color: una tira de 2.8-pies de color amarillo, una tira naranja de 2.08 pies y una tira roja de 2.25 pies.

Colocó la tira más corta en un cajón y colocó las otras 2 tiras lado a lado sobre una mesa. Dibuja un diagrama de cinta de la comparación de las longitudes de las tiras en la mesa. ¿Qué medida es más larga?

_____

_____

_____

_____

Lee   **Dibuja**   **Escribe**

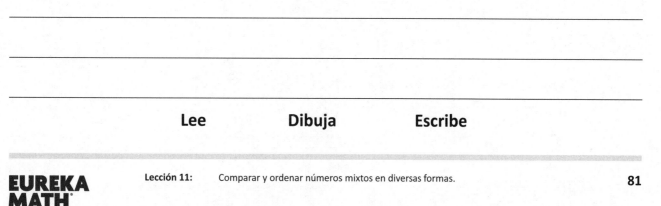

© 2019 Great Minds®. eureka-math.org

Nombre _____ Fecha _____

1. Representa los siguientes puntos en la recta numérica.

    a. 0.2, $\frac{1}{10}$, 0.33, $\frac{12}{100}$, 0.21, $\frac{32}{100}$

    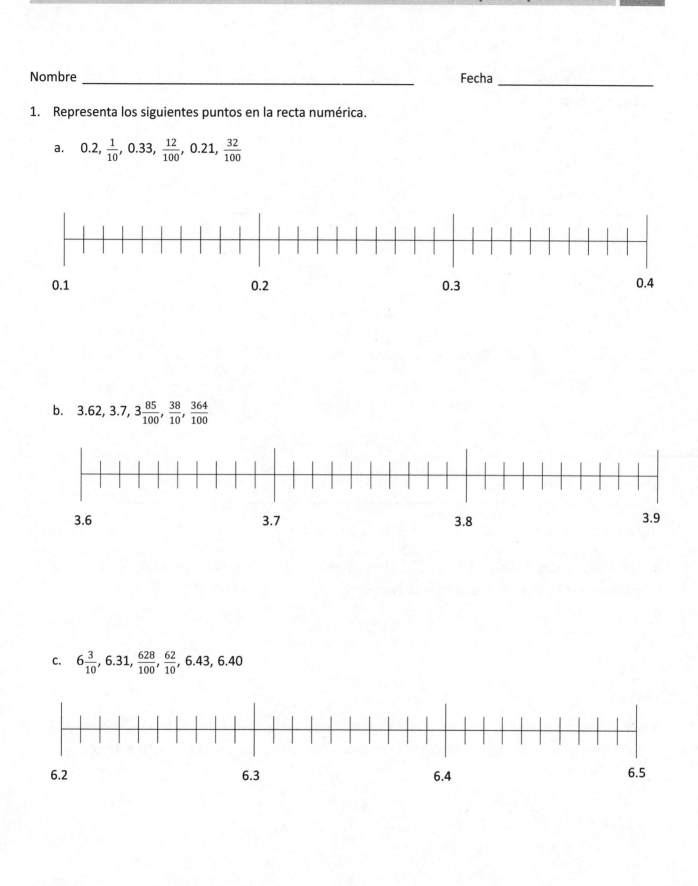

    b. 3.62, 3.7, $3\frac{85}{100}$, $\frac{38}{10}$, $\frac{364}{100}$

    c. $6\frac{3}{10}$, 6.31, $\frac{628}{100}$, $\frac{62}{10}$, 6.43, 6.40

Lección 11:   Comparar y ordenar números mixtos en diversas formas.

© 2019 Great Minds®. eureka-math.org

83

2. Ordena los siguientes números de mayor a menor utilizando la forma decimal. Utiliza el símbolo > entre cada número.

a. $\frac{27}{10}$, 2.07, $\frac{27}{100}$, $2\frac{71}{100}$, $\frac{227}{100}$, 2.72

_____

b. $12\frac{3}{10}$, 13.2, $\frac{134}{100}$, 13.02, $12\frac{20}{100}$

_____

c. $7\frac{34}{100}$, $7\frac{4}{10}$, $7\frac{3}{10}$, $\frac{750}{100}$, 75, 7.2

_____

3. En el caso de salto largo, Rhonda saltó 1.64 metros. María saltó $1\frac{6}{10}$ de metro. Kerri saltó $\frac{94}{100}$ de metro. Michelle saltó 1.06 metros. ¿Quién saltó más lejos?

4. En diciembre, cayeron $2\frac{3}{10}$ pies de nieve. En enero cayeron 2.14 pies de nieve. En febrero, cayeron $2\frac{19}{100}$ pies de nieve, y en marzo, cayeron $1\frac{1}{10}$ pies de nieve. ¿Durante qué mes nevó más? ¿Durante qué mes nevó menos?

© 2019 Great Minds®. eureka-math.org

Nombre _____   Fecha _____

1. Representa los siguientes puntos en la recta numérica utilizando la forma decimal.

   1 unidad y 1 decima, $\frac{13}{10}$, 1 unidad y 20 centesimas, $\frac{129}{100}$, 1.11, $\frac{102}{100}$

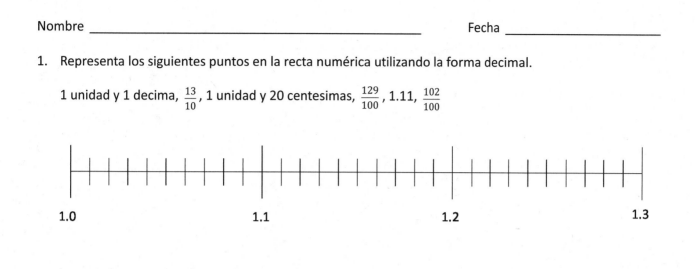

   1.0   1.1   1.2   1.3

2. Ordena los siguientes números de mayor a menor utilizando la forma decimal. Utiliza el símbolo > entre cada número.

   5.6, $\frac{605}{100}$, 6.15, $6\frac{56}{100}$, $\frac{516}{100}$, 6 unidades y 5 décimas

EUREKA MATH®

Lección 11:   Comparar y ordenar números mixtos en diversas formas.

85

© 2019 Great Minds®. eureka-math.org

El lunes cayeron $1\frac{7}{8}$ pulgadas de lluvia. El martes, llovió $\frac{1}{4}$ pulgadas. ¿Cuál fue la precipitación total para los dos días?

_____

_____

_____

Lee   **Dibuja**   **Escribe**

Lección 12:   Aplicar la comprensión de la fracción equivalente para sumar décimas
y centésimas.

87

EUREKA
MATH

© 2019 Great Minds®. eureka-math.org

Nombre _____   Fecha _____

1. Completa el enunciado numérico expresando cada parte usando centésimas. Representa utilizando la tabla de valor posicional, como se muestra en la parte (a).

unidades		décimas	centésimas

a. 1 décima + 5 centésimas = _____ centésimas

unidades		décimas	centésimas

b. 2 décimas + 1 centésima = _____ centésimas

unidades		décimas	centésimas

c. 1 décima + 12 centésimas = _____ centésimas

2. Resuelve mediante la conversión de todos los sumandos a centésimas antes de resolver.

a. 1 décima + 3 centésimas = _____ centésimas + 3 centésimas = _____ centésimas

b. 5 décimas + 12 centésimas = _____ centésimas + _____ centésimas = _____ centésimas

c. 7 décimas + 27 centésimas = _____ centésimas + _____ centésimas = _____ centésimas

d. 37 centésimas + 7 décimas = _____ centésimas + _____ centésimas = _____ centésimas

Lección 12:   Aplicar la comprensión de la fracción equivalente para sumar décimas y centésimas.

89

© 2019 Great Minds®. eureka-math.org

3. Encuentra la suma. Convierte décimas a centésimas según sea necesario. Escribe tu respuesta como decimal.

   a. $\frac{2}{10} + \frac{8}{100}$

   b. $\frac{13}{100} + \frac{4}{10}$

   c. $\frac{6}{10} + \frac{39}{100}$

   d. $\frac{70}{100} + \frac{3}{10}$

4. Resuelve. Escribe tu respuesta como decimal.

   a. $\frac{9}{10} + \frac{42}{100}$

   b. $\frac{70}{100} + \frac{5}{10}$

   c. $\frac{68}{100} + \frac{8}{10}$

   d. $\frac{7}{10} + \frac{87}{100}$

5. El matraz A tiene $\frac{63}{100}$ de litro de yodo. Se llenó el resto con agua hasta 1 litro. El matraz B tiene $\frac{4}{10}$ de litro de yodo. Se llenó el resto con agua hasta 1 litro. Si ambos matraces se vacían en un matraz más grande, ¿cuánto yodo contiene del matraz grande?

Lección 12:   Aplicar la comprensión de la fracción equivalente para sumar décimas y centésimas.

© 2019 Great Minds®. eureka-math.org

Nombre _____   Fecha _____

1. Completa el enunciado numérico expresando cada parte usando centésimas. Usa la tabla de valor posicional para representar.

unidades		décimas	centésimas

1 décima + 9 centésimas = _____ centésimas

2. Encuentra la suma. Escribe tu respuesta como decimal.

$$\frac{4}{10} + \frac{73}{100}$$

Lección 12:   Aplicar la comprensión de la fracción equivalente para sumar décimas y centésimas.

91

EUREKA MATH®

© 2019 Great Minds®. eureka-math.org

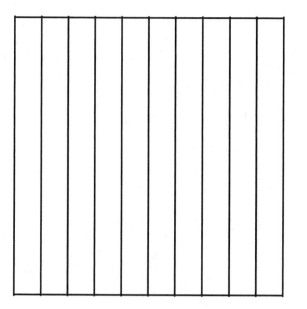

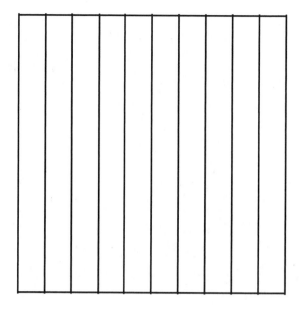

unidades	●	décimas	centésimas

modelo de área y tabla de valor posicional

Lección 12:   Aplicar la comprensión de la fracción equivalente para sumar décimas y centésimas.

93

© 2019 Great Minds®. eureka-math.org

Nombre _____ Fecha _____

1. Resuelve. Convierte de décimas a centésimas antes de encontrar la suma. Reescribe el enunciado numérico completo en forma decimal. Problemas 1 (a) y 1 (b) están resueltos parcialmente.

a. $2\frac{1}{10} + \frac{3}{100} = 2\frac{10}{100} + \frac{3}{100} =$ _____ $2.1 + 0.03 =$ _____	b. $2\frac{1}{10} + 5\frac{3}{100} = 2\frac{10}{100} + 5\frac{3}{100} =$ _____
c. $3\frac{24}{100} + \frac{7}{10}$	d. $3\frac{24}{100} + 8\frac{7}{10}$

2. Resuelve. Luego, reescribe el enunciado numérico completo en forma decimal.

a. $6\frac{9}{10} + 1\frac{10}{100}$	b. $9\frac{9}{10} + 2\frac{45}{100}$
c. $2\frac{4}{10} + 8\frac{90}{100}$	d. $6\frac{37}{100} + 7\frac{7}{10}$

Lección 13:   Sumar números decimales mediante la conversión a forma de fracción.

95

© 2019 Great Minds®. eureka-math.org

3. Resuelve. Reescribe la expresión en forma de fracción. Después de resolver, reescribe el enunciado numérico en forma decimal.

a. 6.4 + 5.3	b. 6.62 + 2.98
c. 2.1 + 0.94	d. 2.1 + 5.94
e. 5.7 + 4.92	f. 5.68 + 4.9
g. 4.8 + 3.27	h. 17.6 + 3.59

Lección 13:   Sumar números decimales mediante la conversión a forma de fracción.

© 2019 Great Minds®. eureka-math.org

Nombre _____   Fecha _____

Resuelve. Reescribe la expresión en forma de fracción. Después de resolver, reescribe el enunciado numérico en forma decimal.

1.   7.3 + 0.95

2.   8.29 + 5.9

Nombre _____   Fecha _____

1. El barril A contiene 2.7 litros de agua. El barril B contiene 3.09 litros de agua. En conjunto, ¿Cuánta agua contienen los dos barriles?

2. Alissa corrió una distancia de 15.8 kilómetros una semana y 17.34 kilómetros la semana siguiente. ¿Qué tanto corrió en las dos semanas?

Lección 14:   Resolver problemas escritos que involucran la suma de medidas en forma decimal.

99

© 2019 Great Minds®. eureka-math.org

3. Un huerto de manzanas vendió 140.5 kilogramos de manzanas en la mañana y 15.85 kilogramos más manzanas por la tarde que por la mañana. ¿Cuántos kilogramos de manzanas se vendieron ese día en total?

4. Un equipo de tres corrió una carrera de relevos. El tiempo del corredor final fue el más rápido, con 29.2 segundos. El tiempo del corredor del medio fue 1.89 segundos más lento que el tiempo del corredor final. El tiempo del corredor inicial fue 0.9 segundos más lento que el tiempo del corredor del medio. ¿Cuál fue el tiempo total del equipo para la carrera?

© 2019 Great Minds®. eureka-math.org

Nombre _____ Fecha _____

Elise corrió 6.43 kilómetros el sábado y 5.6 kilómetros el domingo. ¿Cuántos kilómetros en total corrió el sábado y el domingo?

EUREKA MATH®

Lección 14: Resolver problemas escritos que involucran la suma de medidas en forma decimal.

101

© 2019 Great Minds®. eureka-math.org

Al final del día, Cameron contó el dinero en sus bolsillos. Contó 7 monedas de un centavo, 2 de diez centavos y 2 de veinticino centavos. Di la cantidad de dinero, en centavos, que estaba en los bolsillos de Cameron.

Lee   **Dibuja**   **Escribe**

Lección 15: Expresar cantidades de dinero dadas en diversas formas como números decimales.

103

© 2019 Great Minds®. eureka-math.org

Nombre _____   Fecha _____

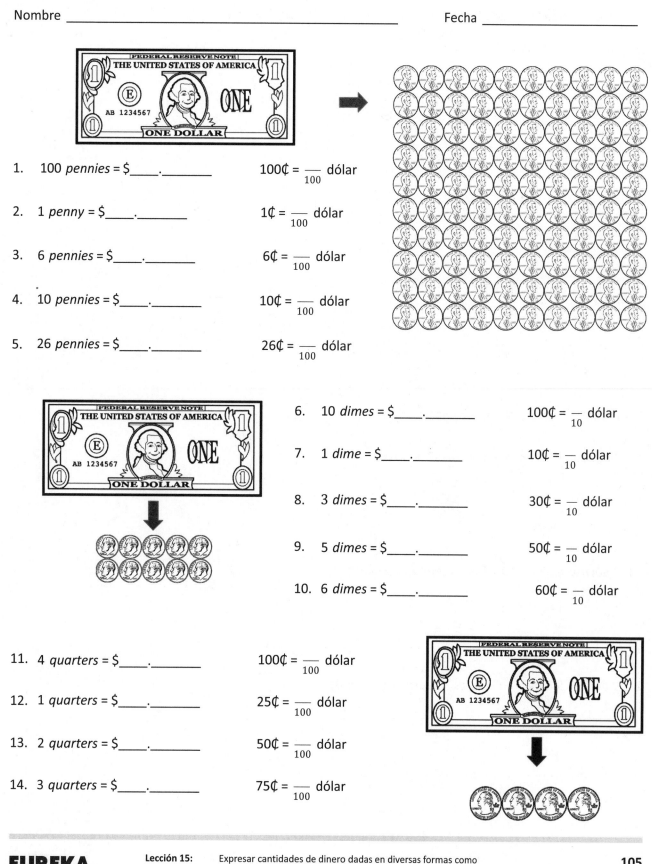

1. 100 *pennies* = $_____._____   100¢ = ——— dólar
                                              100

2. 1 *penny* = $_____._____   1¢ = ——— dólar
                                             100

3. 6 *pennies* = $_____._____   6¢ = ——— dólar
                                             100

4. 10 *pennies* = $_____._____   10¢ = ——— dólar
                                             100

5. 26 *pennies* = $_____._____   26¢ = ——— dólar
                                             100

6. 10 *dimes* = $_____._____   100¢ = ——— dólar
                                              10

7. 1 *dime* = $_____._____   10¢ = ——— dólar
                                             10

8. 3 *dimes* = $_____._____   30¢ = ——— dólar
                                             10

9. 5 *dimes* = $_____._____   50¢ = ——— dólar
                                             10

10. 6 *dimes* = $_____._____   60¢ = ——— dólar
                                             10

11. 4 *quarters* = $_____._____   100¢ = ——— dólar
                                              100

12. 1 *quarters* = $_____._____   25¢ = ——— dólar
                                             100

13. 2 *quarters* = $_____._____   50¢ = ——— dólar
                                             100

14. 3 *quarters* = $_____._____   75¢ = ——— dólar
                                             100

EUREKA MATH

Lección 15:   Expresar cantidades de dinero dadas en diversas formas como
              números decimales.

© 2019 Great Minds®. eureka-math.org

105

Resuelve. Expresa la cantidad total de dinero en forma decimal y fracción.

15. 3 *dimes* y 8 *pennies*.

16. 8 *dimes* y 23 *pennies*.

17. 3 *quarters* 3 *dimes* y 5 *pennies*.

18. ¿Qué fracción de un dólar es 236 centavos?

Resuelve. Expresa la respuesta como un decimal.

19. 2 dólares 17 *pennies* + 4 dólares 2 *quarters*

20. 3 dólares 8 *dimes* + 1 dólar 2 *quarters* 5 *pennies*

21. 9 dólares 9 *dimes* + 4 dólares 3 *quarters* 16 *pennies*

Lección 15:   Expresar cantidades de dinero dadas en diversas formas como números decimales.

© 2019 Great Minds®. eureka-math.org

Nombre _____ Fecha _____

Resuelve. Expresa la cantidad total de dinero en forma decimal y fracción.

1. 2 *quarters* y 3 *dimes*

2. 1 *quarter* 7 *dimes* y 23 *pennies*

Resuelve. Expresa la respuesta como un decimal.

3. 2 dólares 1 *quarter* 14 *pennies* + 3 dólares 2 *quarters* 3 *dimes*

Lección 15:   Expresar cantidades de dinero dadas en diversas formas como
números decimales.

© 2019 Great Minds®. eureka-math.org

107

Nombre _____ Fecha _____

Usa el proceso LDE para resolver los problemas. Escribe tu respuesta como decimal.

1. Miguel tiene 1 billete de un dólar, 2 *dimes* y 7 *pennies*. John tiene 2 billetes de un dólar, 3 *quarters* y 9 *pennies*. ¿Cuánto dinero tienen los dos niños en total?

2. Suilin necesita 7 dólares 13 centavos para comprar un libro. En su cartera, encuentra 3 billetes de un dólar, 4 *dimes* y 14 *pennies*. ¿Cuánto dinero más necesita Suilin para comprar el libro?

3. Vanessa tiene 6 *dimes* y 2 *pennies*. Joachim tiene 1 dólar, 3 *dimes* y 5 *pennies*. Jimmy tiene 5 dólares y 7 *pennies*. Ellos quieren reunir su dinero para comprar un juego que cuesta $ 8.00. ¿Tienen suficiente dinero para comprar el juego? Si no es así, ¿cuánto dinero necesitan?

4. Un bolígrafo cuesta $2.29. Una calculadora cuesta 3 veces más que un bolígrafo. ¿Cuánto cuestan un bolígrafo y una calculadora juntos?

5. Krista tiene 7 dólares y 32 centavos. Malory tiene 2 dólares y 4 centavos. ¿Cuánto dinero necesita dar Krista a Malory para que cada una de ellas tenga la misma cantidad de dinero?

© 2019 Great Minds®. eureka-math.org

Nombre _____   Fecha _____

Usa el proceso LDE para resolver los problemas. Escribe tu respuesta como decimal.

La madre de David le dijo que podría quedarse con todo el dinero que encuentre debajo de los cojines del sofá en su casa. David encuentra 6 *quarters* 4 *dimes* y 26 *pennies*. ¿Cuánto dinero encuentra en total David?

4.° grado

Módulo 7

Nombre _____   Fecha _____

a.

Libras	Onzas
1	
2	
3	
4	
5	
6	
7	
8	
9	
10	

La regla para convertir libras a onzas es _____.

b.

Yardas	pies
1	
2	
3	
4	
5	
6	
7	
8	
9	
10	

La regla para convertir yardas a pies es

_____.

c.

pies	Pulgadas
1	
2	
3	
4	
5	
6	
7	
8	
9	
10	

La regla para convertir pies a pulgadas es

_____.

Lección 1: Crear tablas de conversión para unidades de longitud, peso y capacidad usando
herramientas de medición, y usar las tablas para resolver problemas.

115

© 2019 Great Minds®. eureka-math.org

Nombre _____   Fecha _____

Usa LDE para resolver los Problemas 1–3.

1. Evan puso un peso de 2 libras en un lado de la balanza. ¿Cuántos pesos de 1 onza debe poner en el otro lado de la balanza para equilibrar los pesos?

2. Julius puso un peso de 3 libras en un lado de la balanza. Abel puso 35 pesos de 1 onza en el otro lado. ¿Cuántos pesos más de 1 onza debe poner Abel para equilibrar la balanza?

3. El bebé de la Srta. Upton pesó 5 libras y 4 onzas. ¿Cuántas onzas pesa en total el bebé?

4. Completa las siguientes tablas de conversiones y escribe la regla debajo de cada tabla.

    a.

Libras	Onzas
1	
3	
7	
10	
17	

    La regla para convertir libras a onzas es _____.

Lección 1: Crear tablas de conversión para unidades de longitud, peso y capacidad usando herramientas de medición, y usar las tablas para resolver problemas.

117

© 2019 Great Minds®. eureka-math.org

b.

Pies	Pulgadas
1	
2	
5	
10	
15	

La regla para convertir pies a pulgadas es

_____.

c.

Yardas	Pies
1	
2	
4	
10	
14	

La regla para convertir yardas a pies es

_____.

5. Resuelve.

a. 3 pies 1 pulgada = _____ pulgadas

b. 11 pies 10 pulgadas = _____ pulgadas

c. 5 yardas 1 pie = _____ pies

d. 12 yardas 2 pies = _____ pies

e. 27 libras 10 onzas = _____ onzas

f. 18 yardas 9 pies = _____ pies

g. 14 libras 5 onzas = _____ onzas

h. 5 yardas 2 pies = _____ pulgadas

6. Contesta *verdadero* o *falso* a las siguientes afirmaciones. Si la afirmación es falsa, cambia el lado derecho de la comparación para volverla verdadera.

a. 2 kilogramos > 2,600 gramos   _____

b. 12 pies < 140 pulgadas   _____

c. 10 kilómetros = 10,000 metros   _____

Lección 1:   Crear tablas de conversión para unidades de longitud, peso y capacidad usando herramientas de medición, y usar las tablas para resolver problemas.

© 2019 Great Minds®. eureka-math.org

Nombre _____   Fecha _____

1. Resuelve.

   a. 8 pies = _____ pulgadas

   b. 4 yardas 2 pies = _____ pies

   c. 14 libras 7 onzas = _____ onzas

2. Contesta *verdadero* o *falso* a las siguientes afirmaciones. Si la afirmación es falsa, cambia el lado derecho de la comparación para volverla verdadera.

   a. 3 libras > 60 onzas   _____

   b. 12 yardas > 40 pies   _____

Lección 1:   Crear tablas de conversión para unidades de longitud, peso y capacidad usando herramientas de medición, y usar las tablas para resolver problemas.

© 2019 Great Minds®. eureka-math.org

119

Nombre _____   Fecha _____

a.

Galones	Cuartos de galón
1	
2	
3	
4	
5	
6	
7	
8	
9	
10	

La regla para convertir galones a cuartos de galón es

_____.

b.

Cuartos de galón	Pintas
1	
2	
3	
4	
5	
6	
7	
8	
9	
10	

La regla para convertir cuartos de galón a pintas es

_____.

c.

Pintas	Tazas
1	
2	
3	
4	
5	
6	
7	
8	
9	
10	

La regla para convertir pintas a tazas es_____.

d.   1 galón = _____ pintas

1 cuarto de galón = _____ tazas

1 galón = _____ tazas

Lección 2: Crear tablas de conversión para unidades de longitud, peso y capacidad usando herramientas de medición, y usar las tablas para resolver problemas.

121

© 2019 Great Minds®. eureka-math.org

Nombre _____ Fecha _____

Usa LDE para resolver los Problemas 1–3.

1. Susie tiene 3 cuartos de galón de leche. ¿Cuántas pintas tiene?

2. Kristin tiene 3 galones y 2 cuartos de galón de agua. Alana necesita la misma cantidad de agua, pero solo tiene 8 cuartos de galón. ¿Cuántos cuartos de galón más de agua necesita Alana?

3. Leonard compró 4 litros de jugo de naranja. ¿Cuántos mililitros de jugo tiene?

4. Completa las siguientes tablas de conversiones y escribe la regla debajo de cada tabla.

a.

Galones	Cuartos de galón
1	
3	
5	
10	
13	

La regla para convertir galones a cuartos de galón es

_____ .

b.

Cuartos de galón	Pintas
1	
2	
6	
10	
16	

La regla para convertir cuartos de galón a pintas es

_____ .

Lección 2:   Crear tablas de conversión para unidades de longitud, peso y capacidad usando herramientas de medición, y usar las tablas para resolver problemas.

123

© 2019 Great Minds®. eureka-math.org

5. Resuelve.

a. 8 galones 2 cuartos de galón = _____ cuartos de galón   b.   15 galones 2 cuartos de galón = _____ cuartos de galón

c. 8 cuartos de galón 2 pintas = _____ pintas   d.   12 cuartos de galón 3 pintas = _____ tazas

e. 26 galones 3 cuartos de galón = _____ pintas   f. 32 galones 2 cuartos de galón = _____ tazas

6. Contesta verdadero o falso a las siguientes afirmaciones. Si respondes falso, vuelve verdadera la afirmación.

a. 1 galón > 4 cuartos de galón   _____

b. 5 litros = 5,000 mililitros   _____

c. 15 pintas < 1 galón 1 taza   _____

7. Russel tiene 5 litros de un medicamento. Si se requieren 2 mililitros para preparar 1 dosis, ¿cuántas dosis puede preparar?

8. Cada mes, la familia Moore bebe 16 galones de leche, y la familia Siler bebe 44 cuartos de galón de leche ¿Qué familia bebe más leche cada mes?

9. En el puesto de limonada de Keith se sirvió limonada en vasos con una capacidad de 1 taza. Si él tiene 9 galones de limonada, ¿cuántas tazas puede vender?

© 2019 Great Minds®. eureka-math.org

Nombre _____ Fecha _____

1. Completa la tabla.

Cuartos de galón	Tazas
1	
2	
4	

2. El doctor de Bonnie le recomendó que tomara 2 tazas de leche al día. Si ella compra 3 cuartos de galón de leche, ¿tendrá suficiente leche para que dure 1 semana? Explica cómo lo sabes.

EUREKA MATH®

Lección 2: Crear tablas de conversión para unidades de longitud, peso y capacidad usando herramientas de medición, y usar las tablas para resolver problemas.

125

© 2019 Great Minds®. eureka-math.org

Nombre _____   Fecha _____

a.

de instrucción	Segundos
1	
2	
3	
4	
5	
6	
7	
8	
9	
10	

La regla para convertir minutos a segundos es

_____.

b.

Horas	de instrucción
1	
2	
3	
4	
5	
6	
7	
8	
9	
10	

La regla para convertir horas a minutos es

_____.

c.

Días	Horas
1	
2	
3	
4	
5	
6	
7	
8	
9	
10	

La regla para convertir días a horas es

_____.

Lección 3:   Crear tablas de conversión para unidades de tiempo, y usar las tablas para resolver problemas.

127

© 2019 Great Minds®. eureka-math.org

Nombre _____   Fecha _____

Usa LDE para resolver los Problemas 1–2.

1. Courtney necesita salir de su casa a las 8:00 a.m. Si se despierta a las 6:00 a.m., ¿cuántos minutos tiene para prepararse? Usa la recta numérica para mostrar tu trabajo.

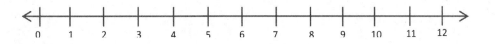

2. El objetivo de Giuliana era correr un maratón en menos de 6 horas. ¿Cuál era su objetivo en minutos?

3. Completa las siguientes tablas de conversiones y escribe la regla debajo de cada tabla.

a.

Horas	de instrucción
1	
3	
6	
10	
15	

b.

Días	Horas
1	
2	
5	
7	
10	

La regla para convertir horas a minutos y minutos a segundos es

_____.

La regla para convertir días a horas es

_____.

Lección 3:   Crear tablas de conversión para unidades de tiempo, y usar las tablas para
              resolver problemas.

129

© 2019 Great Minds®. eureka-math.org

4. Resuelve.

    a. 9 horas 30 minutos = _____ minutos

    b. 7 minutos 45 segundos = _____ segundos

    c. 9 días 20 horas = _____ horas

    d. 22 minutos 27 segundos = _____ segundos

    e. 13 días 19 horas = _____ horas

    f. 23 horas 5 minutos = _____ minutos

5. Explica cómo resolviste el Problema 4(f).

6. ¿Cuántos segundos hay en 14 minutos y 43 segundos?

7. ¿Cuántas horas hay en 4 semanas 3 días?

Lección 3:   Crear tablas de conversión para unidades de tiempo, y usar las tablas para resolver problemas.

© 2019 Great Minds®. eureka-math.org

Nombre _____ Fecha _____

Los astronautas del Apollo 17 completaron 3 caminatas espaciales en la luna con una duración total de 22 horas 4 minutos. ¿Cuántos minutos caminaron los astronautas en el espacio?

Lección 3:   Crear tablas de conversión para unidades de tiempo, y usar las tablas para resolver problemas.

© 2019 Great Minds®. eureka-math.org

131

Nombre _____ Fecha _____

Usa LDE para resolver los siguientes problemas.

1. Beth puede ver 2 horas de televisión cada semana. Su hermana tiene permitido ver el doble. ¿Cuántos minutos de televisión puede ver la hermana de Beth?

2. Clay pesa 9 veces menos que su hermana menor. Clay pesa 63 libras. ¿Cuántas onzas pesa su hermana pequeña?

3. Helen tiene 4 yardas de cuerda. Daniel tiene 4 veces más cuerda que Helen. ¿Cuántos pies más de cuerda tiene Daniel en comparación con Helen?

Lección 4:   Resolver problemas escritos de comparación multiplicativa usando tablas de conversión de medidas.

133

© 2019 Great Minds®. eureka-math.org

4. Un lavavajillas usa 11 litros de agua para cada ciclo. Una lavadora usa 5 veces más agua que la que usa un lavavajillas para cada carga. Juntos, ¿cuántos mililitros de agua se usan para 1 ciclo de cada máquina?

5. Joyce compró 2 libras de manzanas. Ella compró 3 veces más libras de papas que libras de manzanas. Los melones que compró pesaron 10 onzas menos que el peso total de las papas. ¿Cuántas onzas pesaron los melones?

Lección 4: Resolver problemas escritos de comparación multiplicativa usando tablas de conversión de medidas.

© 2019 Great Minds®. eureka-math.org

Nombre _____ Fecha _____

Usa LDE para resolver el siguiente problema.

Brian tiene un melón que pesa 3 libras. Él lo cortó en seis partes iguales. ¿Cuántas onzas pesó cada parte?

Lección 4:   Resolver problemas escritos de comparación multiplicativa usando tablas de conversión de medidas.

135

EUREKA MATH®

© 2019 Great Minds®. eureka-math.org

Nombre _____ Fecha _____

1. a. Etiqueta el resto del siguiente diagrama de cinta. Encuentra la parte desconocida.

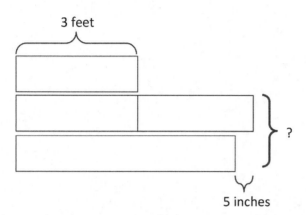

3 feet

?

5 inches

   b. Escribe un problema propio que pueda resolverse usando el diagrama anterior.

2. Crea un problema propio usando el siguiente diagrama y encuentra la incógnita.

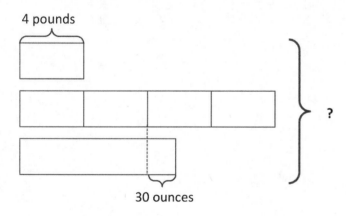

4 pounds

?

30 ounces

Lección 5:   Compartir y criticar las estrategias de los compañeros.

137

© 2019 Great Minds®. eureka-math.org

Nombre _____   Fecha _____

Caitlin corrió 1,680 pies el lunes y 2,340 pies el martes. ¿Cuántas yardas corrió en esos dos días?

Compañero de clase:		Número de problema:	
Estrategias que usó mi compañero de clase:			
Cosas que hizo bien mi compañero de clase:			
Sugerencias para mejorar:			
Cambios que puedo hacer a mi trabajo con base en el trabajo de mi compañero de clase:			

Compañero de clase:		Número de problema:	
Estrategias que usó mi compañero de clase:			
Cosas que hizo bien mi compañero de clase:			
Sugerencias para mejorar:			
Cambios que puedo hacer a mi trabajo con base en el trabajo de mi compañero de clase:			

formulario para analizar y compartir con compañeros

Lección 5:   Compartir y criticar las estrategias de los compañeros.

141

© 2019 Great Minds®. eureka-math.org

Nombre _____   Fecha _____

1. Determina las siguientes sumas y diferencias. Muestra tu trabajo.

    a.   3 qt + 1 qt = _____ gal   b.   2 gal 1 qt + 3 qt = _____ gal

    c.   1 gal − 1 qt = _____ qt   d.   5 gal − 1 qt = _____ gal _____ qt

    e.   2 c + 2 c = _____ qt   f.   1 qt 1 pt + 3 pt = _____ qt

    g.   2 qt − 3 pt = _____ pt   h.   5 qt − 3 c _____ qt _____ c

2. Encuentra las siguientes sumas y diferencias. Muestra tu trabajo.

    a.   6 gal 3 qt + 3 qt = _____ gal _____ qt   b.   10 gal 3 qt + 3 gal 3 qt = _____ gal _____ qt

    c.   9 gal 1 pt − 2 pt = _____ gal _____ pt   d.   7 gal 1 pt − 2 gal 7 pt = _____ gal _____ pt

    e.   16 qt 2 c + 4 c = _____ qt _____ c   f.   6 gal 5 pt + 3 gal 3 pt = _____ gal _____ pt

3. Una jarra tiene una capacidad de 3 cuartos de galón. Ahora, contiene 1 cuarto de galón 3 tazas de líquido. ¿Cuánto líquido más puede contener la jarra?

4. Dorothy sigue la receta en la tabla para hacer la limonada de cereza de su abuela.

    a. ¿Cuánta limonada se hace con la receta?

Limonada de cereza	
Ingrediente	**Cantidad**
Jugo de limón	5 pintas
Jarabe de azúcar	2 tazas
Agua	1 galón 1 cuarto de galón
Jugo de cereza	3 cuartos de galón

    b. ¿Cuántas tazas de agua más podría añadir Dorothy a la receta para preparar un número exacto de galones de limonada?

© 2019 Great Minds®. eureka-math.org

Nombre _____   Fecha _____

1. Encuentra las siguientes sumas y diferencias. Muestra tu trabajo.

    a. 7 gal 2 qt + 3 gal 3 qt = _____ gal _____ qt

    b. 9 gal 1 qt – 5 gal 3 qt = _____ gal _____ qt

2. Jason vertió 1 galón 1 cuarto de galón de agua en una cubeta vacía de 2 galones. ¿Cuánta más agua puede añadirse para llegar a la capacidad de 2 galones de la cubeta?

Lección 6:   Resolver problemas que involucran unidades mixtas de capacidad.

145

© 2019 Great Minds®. eureka-math.org

Samantha va a preparar ponche para un día de campo con la clase. Hay 26 estudiantes en su clase. Samantha usa 1 galón 2 cuartos de galón de jugo de naranja, 3 cuartos de galón de limonada y 1 galón 3 cuartos de galón de agua mineralizada. ¿Cuánto ponche hizo Samantha? ¿Alcanzará para servir a cada estudiante dos raciones de 1 taza de ponche?

_____

_____

_____

_____

Lee   Dibuja   Escribe

EUREKA MATH®

Lección 7:   Resolver problemas que involucran unidades mixtas de longitud.

147

© 2019 Great Minds®. eureka-math.org

Nombre _____   Fecha _____

1. Determina las siguientes sumas y diferencias. Muestra tu trabajo.

    a.   1 pie + 2 pies = _____ yd

    b.   3 yd 1 pie + 2 pies = _____ yd

    c.   1 yd – 1 pie = _____ pie

    d.   8 yd – 1 pie = _____ yd _____ pie

    e.   3 in + 9 in = _____ pie

    f.   6 in + 9 in = _____ pie _____ in

    g.   1 pie – 8 in = _____ in

    h.   5 pies – 8 in = _____ pie _____ in

2. Encuentra las siguientes sumas y diferencias. Muestra tu trabajo.

    a.   5 yd 2 pies + 2 pies = _____ yd _____ pie

    b.   7 yd 2 pies + 2 yd 2 pies = _____ yd _____ pie

    c.   4 yd 1 pies – 2 pies = _____ yd _____ pie

    d.   6 yd 1 pie – 2 yd 2 pies = _____ yd _____ pie

    e.   6 pies 9 in + 4 in = _____ pies _____ in

    f.   4 pies 4 in + 3 pies 11 in = _____ pies _____ in

    g.   34 pies 4 in – 8 in = _____ pies _____ in

    h.   7 pies 1 in – 5 pies 10 in = _____ pies _____ in

3. Matthew mide 6 pies 2 pulgadas de alto. Su prima pequeña, Emma, mide 3 pies 6 pulgadas de alto. ¿Cuánto más alto es Matthew que Emma?

4. En la clase de gimnasia, Jared escaló 10 pies 4 pulgadas sobre una cuerda. Después, siguió escalando otros 3 pies 9 pulgadas. ¿Hasta qué altura escaló Jared?

5. Un cuadrilátero tiene un perímetro de 18 pies 2 pulgadas. La suma de tres de los lados son 12 pies 4 pulgadas.

    a. ¿Cuál es la longitud del cuarto lado?

    b. El lado de un triángulo equilátero tiene una longitud igual al cuarto lado de un cuadrilátero. ¿Cuál es el perímetro del triángulo?

Lección 7:   Resolver problemas que involucran unidades mixtas de longitud.

© 2019 Great Minds®. eureka-math.org

Nombre _____ Fecha _____

Determina las siguientes sumas y restas. Muestra tu trabajo.

1. 4 yd 1 pie + 2 pies _____ yd

2. 6 yd – 1 pie = _____ yd _____ pie

3. 4 yd 1 pie + 3 yd 2 pies = _____ yd

4. 8 yd 1 pie – 3 yd 2 pies = _____ yd _____ pie

Lección 7:   Resolver problemas que involucran unidades mixtas de longitud.

151

© 2019 Great Minds®. eureka-math.org

Un letrero junto a la montaña rusa indica que una persona debe medir 54 pulgadas de altura para subirse. En su última cita médica, Hever midió 4 pies 4 pulgadas de altura. Él ha crecido 3 pulgadas desde entonces.

a. ¿Hever tiene la altura suficiente para subirse a la montaña rusa? ¿Cuántas pulgadas le faltan o le sobran con respecto a la altura mínima?

b. El papá de Hever mide 6 pies 3 pulgadas de altura. ¿Cuánto más alto que la altura mínima es su papá?

Lee   **Dibuja**   **Escribe**

Lección 8:   Resolver problemas que involucran unidades mixtas de peso.

153

© 2019 Great Minds®. eureka-math.org

Nombre _____   Fecha _____

1. Determina las siguientes sumas y diferencias. Muestra tu trabajo.

    a.   7 oz + 9 oz = _____ lb   b.   1 lb 5 oz + 11 oz = _____ lb

    c.   1 lb – 13 oz = _____ oz   d.   12 lb – 4 oz = _____ lb _____ oz

    e.   3 lb 9 oz + 9 oz = _____ lb _____ oz   f.   30 lb 9 oz + 9 lb 9 oz _____ lb _____ oz

    g.   25 lb 2 oz – 14 oz = _____ lb _____ oz   h.   125 lb 2 oz – 12 lb 3 oz = _____ lb _____ oz

2. El peso total de las mochilas llenas de Sarah y Amanda es de 27 libras. La mochila de Sarah pesa 15 libras
    9 onzas. ¿Cuánto pesa la mochila de Amanda?

3. En la cartuchera de Emma, un lápiz pesa 3 onzas. Sus tijeras pesan 3 onzas más que el lápiz, y una botella de pegamento pesa tres veces más que las tijeras. ¿Cuánto pesa la botella de pegamento en libras y onzas?

4. Usa la información en la tabla sobre los útiles escolares de Jodi para contestar las siguientes preguntas:

   a. Los lunes, Jodi solo guarda su laptop y cartuchera en su mochila. ¿Cuánto pesa su mochila llena?

Libro de texto	Cartuchera 1 lb	Carpeta 2 lb 5 oz
Laptop 5 lb 12 oz	Cuaderno 11 oz	Mochila (vacía) 2 lb 14 oz

   b. Los martes, Jodi lleva su laptop, su cartuchera, dos cuadernos y dos libros de texto en su mochila. Los viernes, Jodi solo guarda su carpeta y su cartuchera. ¿Cuánto menos pesa la mochila llena de Jodi los viernes de lo que pesa los martes?

© 2019 Great Minds®. eureka-math.org

Nombre _____   Fecha _____

Determina las siguientes sumas y diferencias. Muestra tu trabajo.

1. 4 lb 6 oz + 10 oz = _____ lb _____ oz

2. 12 lb 4 oz + 3 lb 14 oz = _____ lb _____ oz

3. 5 lb 4 oz – 12 oz = _____ lb _____ oz

4. 20 lb 5 oz – 13 lb 7 oz = _____ lb _____ oz

Lección 8:   Resolver problemas que involucran unidades mixtas de peso.

157

© 2019 Great Minds®. eureka-math.org

Nombre _____ Fecha _____

1. Determina las siguientes sumas y restas. Muestra tu trabajo.

    a.   23 min + 37 min = _____ hr

    b.   1 hr 11 min + 49 min = _____ hr

    c.   1 hr – 12 min = _____ min

    d.   4 hr – 12 min = _____ hr _____ min

    e.   22 seg + 38 seg = _____ min

    f.   3 min – 45 seg = _____ min _____ seg

2. Encuentra las siguientes sumas y restas. Muestra tu trabajo.

    a.   3 hr 45 min + 25 min = _____ hr _____ min

    b.   2 hr 45 min + 6 hr 25 min = _____ hr _____ min

    c.   3 hr 7 min – 42 min = _____ hr _____ min

    d.   5 hr 7 min – 2 hr 13 min = _____ hr _____ min

    e.   5 min 40 seg + 27 seg = _____ min _____ seg

    f.   22 min 48 seg – 5 min 58 seg = _____ min ____ seg

Lección 9:   Resolver problemas que involucran unidades mixtas de tiempo.

159

© 2019 Great Minds®. eureka-math.org

3. En la competencia de apilado de vasos, el tiempo del primer lugar fue 1 minuto 52 segundos. Eso fue 31 segundos más rápido que el segundo lugar. ¿Cuál fue el tiempo del segundo lugar?

4. Jackeline y Raychel tienen 5 horas para ver tres películas que duran 1 hora 22 minutos, 2 horas 12 minutos y 1 hora 57 minutos respectivamente.

    a. ¿Las chicas tienen suficiente tiempo para ver las tres películas? Explica por qué sí o por qué no.

    b. Si Jackeline y Raychel deciden solo ver las dos películas más largas y tomarse un descanso de 30 minutos en el medio, ¿cuánto tiempo les quedará de sus 5 horas?

© 2019 Great Minds®. eureka-math.org

Nombre _____   Fecha _____

Encuentra las siguientes sumas y restas. Muestra tu trabajo.

1.   2 hr 25 min + 25 min = _____ hr _____ min

2.   4 hr 45 min + 2 hr 35 min = _____ hr _____ min

3.   11 hr 6 min – 32 min = _____ hr _____ min

4.   8 hr 9 min – 6 hr 42 min = _____ hr _____ min

Lección 9:   Resolver problemas que involucran unidades mixtas de tiempo.

161

© 2019 Great Minds®. eureka-math.org

Nombre _____ Fecha _____

Usa LDE para resolver los siguientes problemas.

1. El tiempo de natación de Paula en el Triatlón Ironman fue de 1 hora 25 minutos. Su tiempo en ciclismo fue 5 horas mayor que su tiempo en natación. Ella corrió 4 horas 50 minutos. ¿Cuánto tiempo le tomó completar las tres partes de la carrera?

2. Nolan cargó 7 galones y 3 cuartos de galón de gasolina en su automóvil el lunes, y el doble de eso el sábado. ¿Cuál fue la cantidad total de gasolina que cargó en su automóvil en ambos días?

Lección 10: Resolver problemas escritos de medición en varios pasos.

163

© 2019 Great Minds®. eureka-math.org

3. Una calabaza pesa 7 libras 12 onzas. Una segunda calabaza pesa 10 libras 4 onzas. Una tercera calabaza pesa 2 libras 9 onzas más que la segunda calabaza. ¿Cuál es el peso total de las tres calabazas?

4. El Sr. Lane mide 6 pies 4 pulgadas de altura. Su hija, Mary, mide 3 pies 8 pulgadas menos que su papá. Su hijo mide 9 pulgadas más que Mary. ¿Cuántas pulgadas es el Sr. Lane más alto que su hijo?

Lección 10:   Resolver problemas escritos de medición en varios pasos.

© 2019 Great Minds®. eureka-math.org

Nombre _____   Fecha _____

Usa LDE para resolver el siguiente problema.

Hadley pasó 1 hora y 20 minutos completando su tarea de matemáticas, 45 minutos completando su tarea de estudios sociales y 30 minutos estudiando ortografía. ¿Cuánto tiempo pasó Hadley haciendo tarea y estudiando?

EUREKA
MATH®

Lección 10:   Resolver problemas escritos de medición en varios pasos.

165

© 2019 Great Minds®. eureka-math.org

Nombre _____   Fecha _____

Usa LDE para resolver los siguientes problemas.

1. Lauren corrió un maratón y terminó 1 hora 15 minutos después que Amy, quien tuvo un tiempo de 2 horas 20 minutos. Cassie terminó 35 minutos después que Lauren. ¿Cuánto tiempo le tomó a Cassie correr el maratón?

2. El Chef Joe tiene 8 lb 4 oz de carne molida en su congelador. Esto es $\frac{1}{3}$ de la cantidad necesaria para hacer el número de hamburguesas que se planearon para una fiesta. Si usa 4 onzas de carne para cada hamburguesa, ¿cuántas hamburguesas está planeando hacer?

Lección 11:   Resolver problemas escritos de medición en varios pasos.

167

© 2019 Great Minds®. eureka-math.org

3. Sarah leyó durante 1 hora y 17 minutos cada día por 6 días. Si le tomó 3 minutos leer cada página, ¿cuántas páginas leyó en 6 días?

4. Los grados 3, 4 y 5 van juntos a su día de campo anual. A cada grado le entregaron 16 galones de agua. Si hay un total de 350 estudiantes, ¿habrá suficiente agua para que cada estudiante tome 2 tazas?

© 2019 Great Minds®. eureka-math.org

Nombre _____ Fecha _____

Usa LDE para resolver el siguiente problema.

Judy pasó 1 hora y 15 minutos menos que Sandy haciendo ejercicio la semana pasada. Sandy pasó 50 minutos menos que Mary, quien pasó 3 horas en el gimnasio. ¿Cuánto tiempo se ejercitó Judy?

Lección 11:   Resolver problemas escritos de medición en varios pasos.

169

© 2019 Great Minds®. eureka-math.org

Una losa rectangular tiene un ancho de 1 pie 6 pulgadas y un largo de 2 pies. ¿Cuál es el perímetro de la losa?

_____

_____

_____

_____

Lee   **Dibuja**   **Escribe**

Nombre _____ Fecha _____

1. Dibuja un diagrama de cinta para mostrar 1 yarda dividida en 3 partes iguales.

   a. $\frac{1}{3}$ yd = _____ pie

   b. $\frac{2}{3}$ yd = _____ pie

   c. $\frac{3}{3}$ yd = _____ pie

2. Dibuja un diagrama de cinta para mostrar $2\frac{2}{3}$ yardas = 8 pies.

3. Dibuja un diagrama de cinta para mostrar $2\frac{2}{3}$ galón = 3 cuartos de galón.

4. Dibuja un diagrama de cinta para mostrar que $3\frac{3}{4}$ galones = 15 cuartos de galón.

5. Resuelve los problemas usando la herramienta que funcione mejor para ti.

   a. $\frac{1}{12}$ pie = _____ in

   b. $\frac{}{12}$ pie = $\frac{1}{2}$ pie = _____ in

   c. $\frac{}{12}$ pie = $\frac{1}{4}$ pie = _____ in

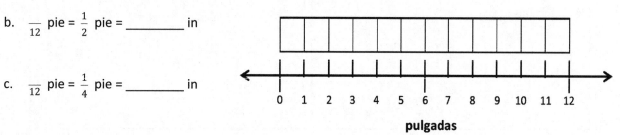

pulgadas

Lección 12:   Usar herramientas de medición para convertir medidas numéricas mixtas en unidades más pequeñas.

173

© 2019 Great Minds®. eureka-math.org

d. $\dfrac{}{12}$ pie = $\dfrac{3}{4}$ pie = _____ in

e. $\dfrac{}{12}$ pie = $\dfrac{1}{3}$ pie = _____ in

f. $\dfrac{}{12}$ pie = $\dfrac{2}{3}$ pie = _____ in

6. Resuelve.

a. $1\dfrac{1}{3}$ yd = _____ pie	b. $4\dfrac{2}{3}$ yd = _____ pie
c. $2\dfrac{1}{2}$ gal = _____ qt	d. $7\dfrac{3}{4}$ gal = _____ qt
e. $1\dfrac{1}{2}$ pie = _____ in	f. $6\dfrac{1}{2}$ pie = _____ in
g. $1\dfrac{1}{4}$ pie = _____ in	h. $6\dfrac{1}{4}$ pie = _____ in

Lección 12: Usar herramientas de medición para convertir medidas numéricas mixtas en unidades más pequeñas.

© 2019 Great Minds®. eureka-math.org

EUREKA
MATH®

Nombre _____   Fecha _____

1. Resuelve los problemas usando la herramienta que funcione mejor para ti.

    a. $\dfrac{}{12}$ pie = $\dfrac{1}{2}$ pie = _____ in

    b. $\dfrac{}{12}$ pie = $\dfrac{3}{4}$ pie = _____ in

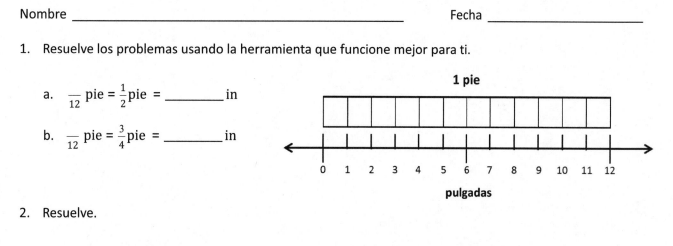

2. Resuelve.

    a. $1\dfrac{1}{3}$ yd = _____ pie

    b. $5\dfrac{3}{4}$ gal = _____ qt

Micah usó $3\frac{3}{4}$ galones de pintura para pintar su baño. Usó 3 veces más pintura para pintar su dormitorio. ¿Cuántos cuartos de galón de pintura necesitó para pintar su dormitorio?

_____

_____

_____

_____

Lee   **Dibuja**   **Escribe**

Lección 13:   Usar herramientas de medición para convertir medidas numéricas mixtas en unidades más pequeñas.

177

© 2019 Great Minds®. eureka-math.org

Nombre _____   Fecha _____

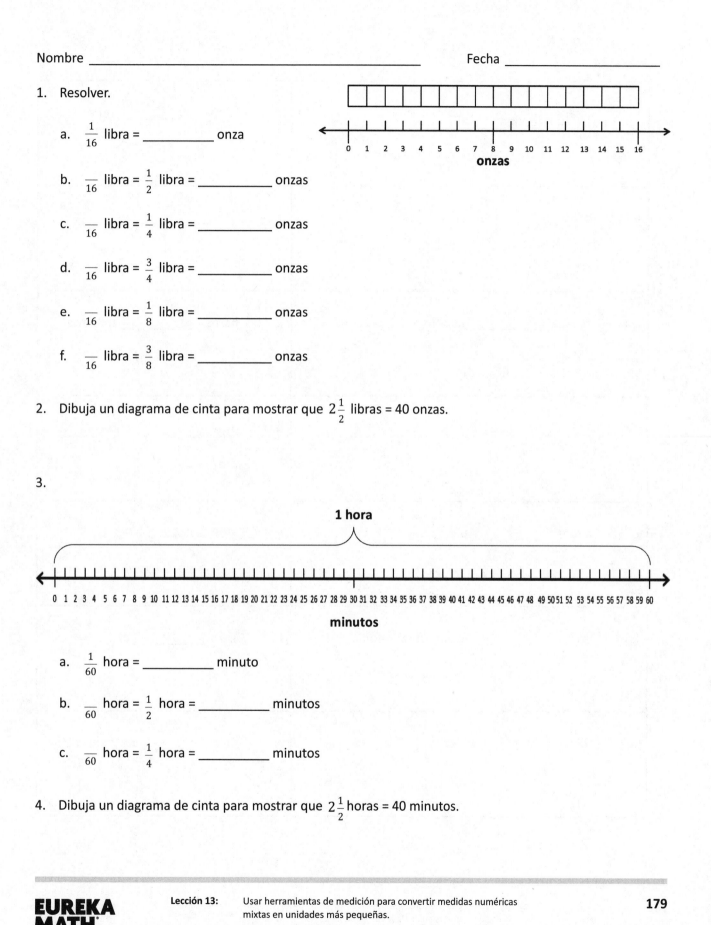

1. Resolver.

     a. $\frac{1}{16}$ libra = _____ onza

     b. $\frac{}{16}$ libra = $\frac{1}{2}$ libra = _____ onzas

     c. $\frac{}{16}$ libra = $\frac{1}{4}$ libra = _____ onzas

     d. $\frac{}{16}$ libra = $\frac{3}{4}$ libra = _____ onzas

     e. $\frac{}{16}$ libra = $\frac{1}{8}$ libra = _____ onzas

     f. $\frac{}{16}$ libra = $\frac{3}{8}$ libra = _____ onzas

2. Dibuja un diagrama de cinta para mostrar que $2\frac{1}{2}$ libras = 40 onzas.

3.

     a. $\frac{1}{60}$ hora = _____ minuto

     b. $\frac{}{60}$ hora = $\frac{1}{2}$ hora = _____ minutos

     c. $\frac{}{60}$ hora = $\frac{1}{4}$ hora = _____ minutos

4. Dibuja un diagrama de cinta para mostrar que $2\frac{1}{2}$ horas = 40 minutos.

© 2019 Great Minds®. eureka-math.org

5. Resuelve.

a. $1\frac{1}{8}$ libras = _____ onzas	b. $3\frac{3}{8}$ libras = _____ onzas
c. $5\frac{3}{4}$ lb = _____ oz	d. $5\frac{1}{2}$ lb = _____ oz
e. $1\frac{1}{4}$ horas = _____ minutos	f. $3\frac{1}{2}$ horas = _____ minutos
g. $2\frac{1}{4}$ hr = _____ min	h. $5\frac{1}{2}$ hr = _____ min
i. $3\frac{1}{3}$ yardas = _____ pies	j. $7\frac{2}{3}$ yd = _____ pie
k. $4\frac{1}{2}$ galones = _____ cuartos de galón	l. $6\frac{3}{4}$ gal = _____ qt
m. $5\frac{3}{4}$ pies = _____ pulgadas	n. $8\frac{1}{3}$ ft = _____ pulgadas

Lección 13: Usar herramientas de medición para convertir medidas numéricas mixtas en unidades más pequeñas.

© 2019 Great Minds®. eureka-math.org

Nombre _____   Fecha _____

1. Dibuja un diagrama de cinta para mostrar que $4\frac{4}{3}$ galones = 19 cuartos de galón.

2. Resuelve.

a. $1\frac{1}{4}$ libras = _____ onzas	b. $2\frac{3}{4}$ hr = _____ min
c. $5\frac{1}{2}$ pies = _____ pulgadas	d. $3\frac{5}{6}$ ft = _____ pulgadas

Lección 13:   Usar herramientas de medición para convertir medidas numéricas
mixtas en unidades más pequeñas.

181

© 2019 Great Minds®. eureka-math.org

Nombre _____ Fecha _____

Usa LDE para resolver los siguientes problemas.

1. Una caricatura dura $\frac{1}{2}$ hora. Una película dura 6 veces más que la caricatura. ¿Cuántos minutos tomará ver la caricatura y la película?

2. Una banca grande mide $7\frac{1}{6}$ pies de largo. Es 17 pulgadas más larga que una banca más corta. ¿Cuántas pulgadas mide de largo la banca más corta?

3. El primer contenedor puede contener 4 galones 2 cuartos de galón de jugo. El segundo contenedor puede contener $\frac{3}{4}$ galones más que el primer contenedor. ¿Cuánto jugo pueden contener los dos contenedores juntos?

Lección 14:   Resolver problemas escritos de varios pasos que involucran convertir
             medidas numéricas mixtas en una sola unidad.

183

© 2019 Great Minds®. eureka-math.org

4. Una niña mide $3\frac{1}{3}$ pies de altura. Una jirafa mide 3 veces lo que mide la niña. ¿Cuántas pulgadas la jirafa es más alta que la niña?

5. Se colocan cinco onzas de pretzels en cada bolsa. ¿Cuántas bolsas pueden hacerse con $22\frac{3}{4}$ libras de pretzels?

6. Veinte porciones of panqueques requieren 15 onzas de mezcla para panqueques.

   a. ¿Cuánta mezcla para panqueques se necesita para 120 porciones?

   b. Extensión: La mezcla se compra en bolsas de $2\frac{1}{2}$ libras. ¿Cuántas bolsas se necesitarán para hacer 120 porciones?

Lección 14:   Resolver problemas escritos de varios pasos que involucran convertir medidas numéricas mixtas en una sola unidad.

© 2019 Great Minds®. eureka-math.org

Nombre _____ Fecha _____

Usa LDE para resolver el siguiente problema.

A Gigi le tomó 1 hora y 20 minutos completar una carrera en bicicleta. A Johnny le tomó el doble de tiempo porque se le ponchó una llanta. ¿Cuántos minutos le tomó a Johnny terminar la carrera?

Lección 14: Resolver problemas escritos de varios pasos que involucran convertir medidas numéricas mixtas en una sola unidad.

185

© 2019 Great Minds®. eureka-math.org

El dormitorio rectangular de Emma tiene 11 pies de largo y 12 pies de ancho. Dibuja e identifica un diagrama del dormitorio de Emma. ¿Cuántos pies cuadrados de alfombra necesita Emma para cubrir el piso de su dormitorio?

_____

_____

_____

_____

Lee   **Dibuja**   **Escribe**

Lección 15:   Crear y determinar el área de figuras compuestas.

187

EUREKA MATH®

© 2019 Great Minds®. eureka-math.org

Nombre _____   Fecha _____

1. El dormitorio rectangular de Emma tiene 11 pies de largo y 12 pies de ancho, con un ropero de 4 pies por 5 pies. ¿Cuántos pies cuadrados de alfombra necesita Emma para cubrir el dormitorio y el ropero?

2. Para ahorrar dinero, Emma ya no va a alfombrar su ropero. Además, quiere que una esquina de 3 por 6 pies de su dormitorio tenga piso de madera. ¿Cuántos pies cuadrados de alfombra necesitará ahora para el dormitorio?

Lección 15:   Crear y determinar el área de figuras compuestas.

189

© 2019 Great Minds®. eureka-math.org

3. Encuentra el área de la imagen que se muestra a la derecha.

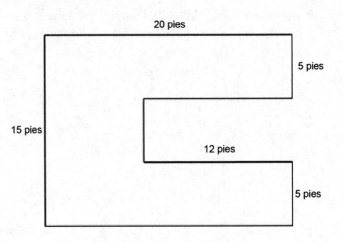

20 pies

5 pies

15 pies

12 pies

5 pies

4. Nombra los lados de la siguiente figura con medidas que tengan sentido. Encuentra el área de la figura.

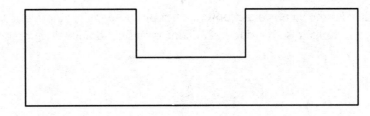

© 2019 Great Minds®. eureka-math.org

EUREKA MATH

5. En el Parque Peterkin hay una fuente con un andador a su alrededor. La fuente mide 12 pies de cada lado. El andador tiene $3\frac{1}{2}$ pies de ancho. Encuentra el área del andador.

6. Si 1 bolsa de grava cubre 9 pies cuadrados, ¿cuántas bolsas de grava se necesitarán para cubrir todo el andador alrededor de la fuente en el Parque Peterkin?

© 2019 Great Minds®. eureka-math.org

Nombre _____   Fecha _____

En la siguiente tabla hay temas que aprendiste en el 4.º Grado y que se usaron en la lección de hoy.

Escoge 1 tema y describe cómo tuviste éxito usándolo hoy.

multiplicación de 2 dígitos por 2 dígitos	Fórmula del área	División de un número de 3 dígitos entre un número de 1 dígito
Resta de números de varios dígitos	Suma de números de varios dígitos	Resolver problemas escritos de varios pasos.

Lección 15:   Crear y determinar el área de figuras compuestas.

193

© 2019 Great Minds®. eureka-math.org

Nombre _____   Fecha _____

Trabajen con su compañero para crear cada plano en una hoja de papel, tal como se describe a continuación.

Deben usar un transportador y una regla para crear cada plano, y asegurarse de que cada rectángulo que creen tenga dos conjuntos de líneas paralelas y cuatro ángulos rectos.

Asegúrense de marcar cada parte de su modelo con la medida correcta.

1. El dormitorio en la casa de muñecas de Samantha es un rectángulo con 26 centímetros de largo y 15 centímetros de ancho. Tiene una cama rectangular que tiene 9 centímetros de largo y 6 centímetros de ancho. Los dos vestidores en el cuarto tienen 2 centímetros de ancho. Uno mide 7 centímetros de largo y el otro mide 4 centímetros de largo. Crea un plano del dormitorio que contenga la cama y los vestidores. Encuentra el área del espacio de piso abierto en el dormitorio después de colocar los muebles.

2. El modelo de una alberca rectangular mide 15 centímetros de largo y 10 centímetros de ancho. La pasarela alrededor de la alberca es 5 centímetros más ancha que la alberca en cada uno de los cuatro lados. En una sección de la pasarela, hay una jardinera que mide 3 centímetros por 5 centímetros. Crea un diagrama del área de la alberca con la pasarela y la jardinera a su alrededor. Encuentra el área abierta de la pasarela alrededor de la alberca.

Lección 16:   Crear y determinar el área de figuras compuestas.

195

© 2019 Great Minds®. eureka-math.org

Nombre _____   Fecha _____

En la siguiente tabla hay habilidades que aprendiste en el 4.º grado y que usaste para completar la lección de hoy. Estas habilidades se introdujeron originalmente en grados anteriores, y seguirás trabajando con ellas durante los siguientes grados. Elige tres temas de la tabla y explica cómo piensas que podrías avanzar y usarlas en el 5to grado.

Multiplicar números de 2 dígitos por 2 dígitos	Usar la fórmula de área para encontrar el área de figuras compuestas	Crear figuras compuestas a partir de un conjunto de especificaciones
Restar números de varios dígitos	Sumar números de varios dígitos	Resolver problemas escritos de varios pasos
Crear líneas paralelas y perpendiculares	Medir y construir Ángulos de 90°	Medir en centímetros

Nombre _____   Fecha _____

1. ¿Qué puedes hacer ahora en matemáticas que no podías hacer al inicio del 4.º grado?

2. ¿Qué actividades te gustaría practicar este verano para mantener la fluidez o desarrollar una mayor fluidez?

3. ¿Qué tipo de práctica te ayudaría a desarrollar una mayor fluidez con estos conceptos?

© 2019 Great Minds®. eureka-math.org

Nombre _____ Fecha _____

1. ¿Por qué piensas que el vocabulario fue una parte tan importante de las matemáticas del cuarto grado? ¿De qué forma te ayuda el vocabulario en matemáticas?

2. ¿Qué términos del vocabulario conoces bien y cuáles te gustaría mejorar?

Lección 18:   Practicar y afianzar la fluidez del vocabulario del 4.º grado.

201

© 2019 Great Minds®. eureka-math.org

Créditos

Great Minds® ha hecho todos los esfuerzos para obtener permisos para la reimpresión de todo el material protegido por derechos de autor. Si algún propietario de material sujeto a derechos de autor no ha sido mencionado, favor ponerse en contacto con Great Minds para su debida mención en todas las ediciones y reimpresiones futuras.

The MIT Press   Cambridge, Massachusetts   London, England

Published in association with the MIT Museum

NIGHTWORK

A HISTORY OF HACKS AND PRANKS AT MIT updated edition

Institute Historian T. F. Peterson ▪ with a new essay by Eric Bender

© 2011 Massachusetts Institute of Technology

All rights reserved. No part of this book may be reproduced in any form by any electronic or mechanical means (including photocopying, recording, or information storage and retrieval) without permission in writing from the publisher.

For information about special quantity discounts, please email special_sales@mitpress.mit.edu
This book was set in Minion and Helvetica Neue Condensed by the MIT Press. Printed and bound in the United States of America.

Library of Congress Cataloging-in-Publication Data

Peterson, T. F.
Nightwork : a history of hacks and pranks at MIT / T.F. Peterson ; with a new essay by Eric Bender.
   p.   cm.
 "Published in association with the MIT Museum."
 Includes index.
  ISBN 978-0-262-51584-9 (pbk. : alk. paper) 1. Massachusetts Institute of Technology—History. 2. Student activities—Massachusetts—Cambridge—History. 3. College students—Massachusetts—Cambridge—Humor. 4. College wit and humor. I. MIT Museum. II. Title.
T171.M49P48   2011
378.744'4—dc22

2010032678

10   9   8   7   6   5   4   3   2   1

CONTENTS

FOREWORD

DEBORAH DOUGLAS, CURATOR OF SCIENCE AND TECHNOLOGY, MIT MUSEUM

My first hack experience was the Harvard–Yale Football Game Hack in 1982. I was a junior at Wellesley College and my father and I were attending the big game. My dad had been part of the Harvard Band, so this was to be an afternoon of father–daughter bonding. More than a quarter-century later, we both still love to tell the story of what suddenly emerged on the field midway through the second quarter! In my wildest imagination back then, I could not have predicted that one day I would be the keeper of that famous hack.

That day would come in 1999 when I became the first curator of science and technology for the MIT Museum. I cannot say for sure, but I suspect it was at least a bit intriguing to my future colleagues when I began my cover letter with the statement that it was 529,894 Smoots from my Virginia office at NASA to the MIT Museum. We get to care for world-class collections at the MIT Museum, but the hacks collection presents some of the most interesting curatorial challenges and enriching conversations to be found at any museum in the world. In museum-speak the word "ephemera" is often used to categorize inconsequential materials; MIT hacks are ephemeral but these fragile constructs are arguably the best record of MIT's soul and student experience for the past half-century. In this special year for MIT, the 150th anniversary of the Institute's founding, I commend this irreverent guide to all who wish to know something of the essence of MIT.

There are many to acknowledge, beginning with the MIT hacking community, which is due incredible heartfelt thanks for pushing the limits and making everyone's experience of MIT memorable. There are a myriad of individuals who generously granted permission to reprint their photographs. Our thanks to you all, but special gratitude to Theresa Smith for helping to track you down! Likewise, we are in debt to Brian Liebowitz, Terri Iuzzolino Matsakis, Stephen Eschenbach, Christopher Pentacoff, and other anonymous consultants who helped polish this imperfect (still!) collection a bit more. They get credit for all the improvements; I eagerly await your corrections and emendations. This is the museum's fourth volume on hacking, and we are grateful to writer and editor Eric Bender for stepping in at the eleventh hour to make this revised edition a reality. Finally, to the saints at the MIT Press, thank you for patience and a willingness to make your daywork be nightwork, too, in order to bring out this new edition as part of the great MIT sesquicentennial celebration.

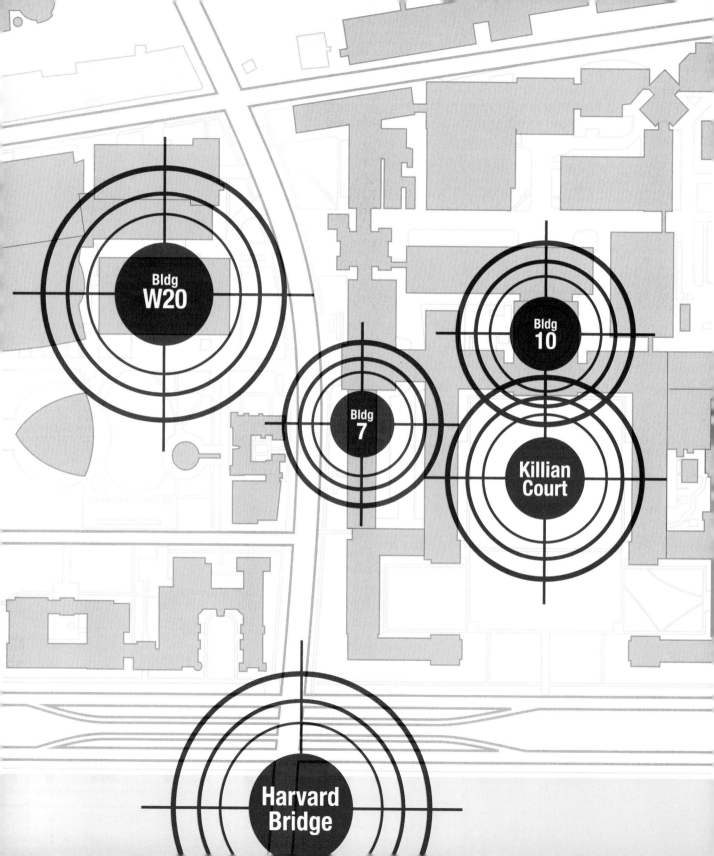

Bldg 32

Bldg 54

Bldg E16

Popular Campus Hack Sites

W20 Stratton Student Center
7 Rogers Building
10 Maclaurin Building
32 Ray and Maria Stata Center
54 Green Building
E15 Wiesner Building
Harvard Bridge
Killian Court

INTRODUCTION/HACKING 1.000H

INSTITUTE HISTORIAN T. F. PETERSON

In 1914, MIT chose the beaver as its mascot from the pages of *Mr. Hornaday's Book on the Animals of North America.* Lester Gardner (Class of 1898) explained why its candidacy was uncontested: "Of all the animals in the world," he said, reading straight from Hornaday, "the beaver is noted for his engineering and mechanical skills and habits of industry. [He is] nocturnal, he does his best work in the dark."

Regardless of superficial changes to campus culture, such as the introduction of computers or mobile phones, the MIT animal has remained true to Hornaday's description at the turn of the twentieth century. The MIT student eats, thinks, daydreams, and socializes under the light of the silvery moon. The black canvas of night is the stimulus for invention. And thus most of the work of hacking—brainstorming, strategy sessions, preparations, test runs, and implementation—happens at night. They may be unveiled by the light of day but, more often than not, they are created at night.

Of course, MIT hackers also emulate the sheer ingenuity of the school mascot. An MIT hack, like science and technology themselves, is judged by how elegantly it accomplishes its objective. The hacking hat trick at the 1982 Harvard–Yale football game, for example, is to many hacking connoisseurs quintessential in its elegance. It reflected the preparation, efficiency, and whimsy that all the most venerable hacks display. More than that, it was a charmingly self-aware, even self-deprecatory statement—another characteristic of the most effective hacks. It said, "Sure, it may be laughable that we could win a football game by our athletic prowess, but we definitely can win it with our brains." And although they took home no trophies, win they did—with an enormous black MIT balloon at the 46-yard line.

In the quarter-century since the balloon burst from the turf at Harvard Stadium, hacking has matured along with the wider culture. When the fraternity responsible for the Harvard–Yale Game hack later admitted its role in the prank, it broke a cardinal rule of hacking culture, a rule that was stringently observed at the time: Hacking was about stealth, not self-aggrandizement. It is still about quiet determination, but unlike a certain modest but enterprising woodland animal, hackers today are a little less modest.

The standards of contemporary hacking etiquette decree that it is also uncouth to create a hack that leaves so much as a disfiguring mark on its environment, and certainly the hack disturbed a bit of turf at the Harvard–Yale game. By the mid-nineties, the challenge of pulling off a great hack included sensitivity to environmental impact. Although the 1994 Entertainment and Hacking frieze required the Institute's Confined Space Rescue Team to rappel down a sheer wall to remove the hack, the team did so with the help of step-by-step dismantling instructions that the hackers delivered to campus authorities.

One key tenet of the sport remains unchanged: The success of a hack is almost directly proportional to the strength of its finer points. The Campus Police Car on the Dome, another legendary 1994 stunt, remains the example par excellence. Passersby began to notice it as dawn broke, but many campus authorities first heard about the hack on metro-Boston traffic reports. Positioned on the Great Dome, as if atop a grand auto showroom dais, sat the shell of a Chevy Cavalier painted to look like a campus police car, its roof lights flashing. A dummy dressed as a police office sat behind the wheel with a half-eaten box of donuts. The car was number π and bore a parking ticket with the offense "no permit for this location."

The Campus Police Car hack is a striking example of how pranks at MIT differ from those at other schools. As Jay Keyser points out in his thought-provoking "Where the Sun Shines, There Hack They" (one of several classic essays on hacking that we have fondly reproduced once again in this edition) MIT students typically don't dress statues of founding fathers in ladies' underwear. They make large objects appear in inaccessible places, rewire lecture hall blackboards, or place a police car or a replica of the Apollo lunar module atop the Dome. "They make fun of engineering," Keyser says, "by impersonating it and then pulling the seat out from under."

But Keyser, as you will read, would be the last to dismiss hacking as a frivolous manifestation of the Institute's engineering culture. For some MIT students, hacking is part and parcel of an MIT education. It teaches them to work productively in teams, to solve engineering problems, and to communicate to the wider world. In "Mastery over the Physical World," André DeHon (1991 SB, 1993 SM, 1996 PhD) discusses how hacking reflects the Institute's own value system. "Hacks provide an opportunity to demonstrate creativity and know-how in mastering the physical world," DeHon observes. "At MIT, intellect and its applications are valued and not, for example, athletic prowess. It's not that we can run faster than you can. It's that we can manipulate the physical world to do things you hadn't imagined were possible."

In its latest incarnation, this book once again pursues the culture of hacking from a myriad of angles, not unlike how hackers themselves work. We will begin with hacking in the past decade and then delve deep into its semi-apocryphal beginnings.

Herewith, we set you loose onto the playing fields of hackdom.

HACK, HACKER, HACKING

The word "hack" at MIT usually refers to a clever, benign, and "ethical" prank or practical joke, which is both challenging for the perpetrators and amusing to the MIT community (and sometimes even the rest of the world!). Note that this has nothing to do with computer (or phone) hacking (which we call "cracking").
—Definition of "hack" from the IHTFP Gallery

A SHORT HISTORY OF THE TERMINOLOGY
Brian Leibowitz, (1982 SB, 1984 SM) author of *The Journal of the Institute for Hacks, TomFoolerly & Pranks at MIT*

The fifties saw the beginnings of the MIT term "hack." The origin of the term in the MIT slang is elusive—different meanings have come in and out of use, and it was rarely used in print before the 1970s. Furthermore, the use of "hack" varied among different groups of students at MIT. "Hacking" was used by many MIT students to describe any activity undertaken to avoid studying—this could include goofing off, playing bridge, talking to friends, or going out. Performing pranks was also considered hacking, but only as part of the broader definition. In the middle to late 1950s, additional meanings for "hack" were developed by members of the Tech Model Railroad Club, including an article or project without constructive end or an unusual and original solution to a problem, such as inventing a new circuit for a switching system. In the late 1950s, students on campus began to use the word as a noun to describe a prank.

Also in the late 1950s, telephone hacking, the study of the internal codes and features of telephone switching systems, emerged. Here, the word "hack" was used to imply doing something outside the norm; telephone systems were made to do things that the system designers never anticipated.

In the late 1960s and the 1970s, the meaning of "hack" broadened to include activities that tested limits of skill, imagination, and wits. Hacking was investigating a subject for its own sake and not for academic advancement, exploring the inaccessible places on campus, doing something clandestine or out of the ordinary, and performing pranks.

The word "hack" found its way into common usage outside MIT with the advent of computer hacking in the early 1960s. In the 1980s, experts in the computer field made a distinction between hacking and cracking. "Hacking" denotes nondestructive mischief while "cracking" describes activities such as unleashing a computer virus, breaking into a computer, or destroying data.

By the mid-1980s, "hacking" had come to be used at MIT primarily to describe pranks and "unapproved exploring" of the Institute. Many of the earlier definitions have disappeared from use on campus.

HACKING ETHICS

In the twenty-first century, the definition of an "ethical hack" is simple. A hack must:
- be safe;
- not damage anything;
- not damage anyone, either physically, mentally, or emotionally; and
- be funny, at least to most of the people who experience it.

For at least a half-century, hacking dogma has been discussed endlessly by MIT students. A good summary of these discussions can be found in the irreverent annual *HowToGAMIT* guide (How to Get around MIT), the ultimate MIT handbook. The first edition in 1969 included a definition of "hacking" but since the 1980s, hacking has been a full feature including a brief historical overview, lists of the most well-known hacking groups, and basic tips. Below is the Hackers' Code of Conduct as present on a mural within the Tomb of The Unknown Tool (1995) as quoted in all recent editions of *HowToGAMIT*.

- The safety of yourself, of others, and of property should have highest priority. Safety is more important than pulling off a hack or getting through a door.
- Be subtle; leave no evidence that you were there.
- Brute force is the last resort of the incompetent.
- Leave things as you found them or better. Cause no permanent damage during hacks and while hacking. If you find something broken call F-IXIT (the campus telephone number for reporting problems with the buildings and grounds).
- Do not steal anything; if you must borrow something, leave a note saying when it will be returned and remember to return it.
- Do not drop things without a ground crew to ensure that no one is underneath.
- Sign-ins are not graffiti and shouldn't be seen by the general public. Sign-ins exhibit one's pride in having found an interesting location and should be seen only by other hackers. Real hackers

are not proud of discovering Lobby 7, random basements, or restrooms. Keep sign-ins small and respect other hackers' sign-ins.

- Never drink and hack.
- Never hack alone. Have someone who can get help in an emergency.
- Know your limitations and do not surpass them. If you do not know how to open a door, climb a shaft, etc., then learn from someone who knows before trying.
- Learn how not to get caught; but if you do get caught, accept gracefully and cooperate fully.
- Share your knowledge and experience with other hackers.
- Above all, exercise common sense.

HACKING INTO THE NEW MILLENNIUM

ERIC BENDER

On the sunny morning of April 6, 2006, a cannon appeared in the courtyard south of MIT's Green Building.

This was not a toy but a two-ton, fully working nineteenth-century weapon.

Closer inspection revealed that its barrel now featured a giant gold-plated version of the Brass Rat, MIT's class ring. An accompanying plaque noted that the cannon was poised to lob shells toward Pasadena, California—where *just such a cannon* had gone missing from the California Institute of Technology nine days earlier. It was, in fact, the famed cannon that normally sits in front of Caltech's Fleming House, and its transcontinental capture drew international media attention once again to MIT's proud tradition of the hack.

"After decades of rivalry between America's geekiest colleges, a group of undergraduates from the Massachusetts Institute of Technology has pulled off one of the most audacious student pranks in history," as the *London Times* put it.

"It was great on many levels," Bill Andrews (2005 SB) later commented in *The Tech*, MIT's student newspaper. "Not only did we show the world how much better MIT really was, not only did they have to come back with their tails between their legs to collect the dumb thing (and rudely refuse the breakfast prepared for them), but on top of all that there was a great picture taken with lots of MIT girls in bikinis posing on it."

"This is the first '21st-century' hack," said Deborah Douglas, curator of science and technology at the MIT Museum. "Its style and mode of organization—a very large group, highly organized, with careful planning, subcontracting work, and an interest in public relations—make it qualitatively different from earlier hacks."

HACKER AND HACKED

Arguably, and you *can* find arguments on campus about this, the Caltech cannon capture was part of a genuine hacking duel.

During MIT's 2005 Campus Preview Weekend, the turbocharged moment each spring when accepted students visit and choose between MIT (and other places including, well, Caltech), MIT itself was hacked. Hacked, you could even say, with style. In among the frolicking and the

Originally built for the Franco-Prussian war, the Caltech cannon has never fired a shot in anger, with the possible rumored exception of potshots toward Caltech's student housing office. In the late '60s, the cannon was "decorating" the campus of a private high school near Caltech, when Fleming House students chopped its wheels out of concrete and swiped it. After a back-and-forth, the cannon ended up permanently at Caltech, except for a brief abduction in 1986 by Harvey Mudd College pranksters and its legendary 2006 cross-country jaunt to MIT.

The giant Brass Rat on the Caltech Cannon took approximately 1,000 hours of machining to create, 2006.

liquid-nitrogen ice cream giveaway and the Mr./Ms. MIT Competition, Caltech students were handing out t-shirts that said "MIT" on the front and " . . . because not everyone can go to Caltech" on the back. They scattered inflatable palm trees at key locations. They stretched a "That Other Institute of Technology" banner across the Massachusetts Avenue face of Building 7 (MIT hackers promptly switched it to say "The Only Institute of Technology"). They even spelled out "Caltech" in green laser light on the Green Building at night. And on a custom Web site, they proposed that MIT join them in a pranking/hacking war.

Duly noted.

A hacking document for MIT's retaliation (yes, some hacking documents *do* survive) shows that the idea of bringing the Fleming or Caltech cannon to MIT dates back at least to 1993. Some West Campus students had begun kicking around the idea in the summer of 2004, almost a full year before the Caltech CPW visit. Now with the added impetus of the Caltech prank, a full-blown plan for the heist emerged by December 2005.

The scheme didn't lack ambition. The Howe & Ser hack team (a pun on "howitzer") eventually included about two dozen people from all over campus. Some helped to craft the giant ring, some fetched the cannon, and a few did both by sacrificing sleep for a month. Trucking the cannon east required hiring a mover—the cover story was that the cannon was going to be used in a movie—which required substantial funding, so the hackers launched a suitably discrete and vaguely worded funding campaign via email.

Approximately a thousand hours of work went into the cannon's Brass Rat, the most durable of any hack artifact ever made, which now resides in the MIT Museum. The hackers told the MIT machinists who assisted in its creation that the pieces were components for a solar-powered motorcycle.

During spring break, five Howe & Ser hackers drove out to California and met up with a few more of their colleagues. Casing out the scene, they also built a wooden frame to hold the cannon for its long journey. Around 5 a.m. on March 28, 2006, they struck, quickly hitching the cannon to their pickup truck. (As good luck had it, the cannon had been moved to a temporary location not directly under the eyes of Flems, as Fleming House residents are known.) Stopped by security guards, they flourished a forged work order and were let go. Twenty minutes later, the campus security office jumped into action after a call reported that the cannon was rolling along and outward bound, but their search proved fruitless.

Safely off campus, the hackers hit another snag when the commercial mover balked at the cannon's heft. The team quickly lined up a second firm, and the cannon began its long journey east the following day. Arriving at MIT on March 31, it was hidden well.

At first, Caltech administrators thought the disappearance was probably a repeat of Harvey Mudd College's kidnapping of the cannon twenty years earlier, almost to the day. When Harvey Mudd denied it and no one else claimed responsibility, Caltech began to worry and filed a grand theft report. After that news spread, Caltech's head of security received an anonymous phone call assuring him that the cannon was safe and would soon be revealed.

The great reveal came on April 6, the Thursday before MIT's Campus Preview Weekend. The Howe & Ser website began a brisk business in T-shirts. The world's media descended on the cannon, as did tours of prefrosh. The "Cannon College Co-Eds" bikini team posed for pictures, which one Caltech professor noted may have done more to demoralize Fleming House students than the cannon abduction itself.

Most students at Caltech found the stunt amusing, as did the administration. Caltech president (and former MIT Institute Professor) David Baltimore even called it an "imaginative response."

Fleming House mobilized to retake their cannon, with twenty-three students flying out to Boston, where they were joined by seven alums. The cannon being well-guarded around the clock, the Caltech crew abandoned their ambitious rescue plans (including a helicopter lift!).

Near dawn on April 9, "MIT students played Wagner's 'Ride of the Valkyries' on a stereo as the Flems emerged from Building 66 and prepared the cannon for transportation, then rolled it to Ames Street, where a flatbed truck was waiting," *The Tech* reported. "Afterward, students from both schools joined together on the Dot [a large circular grassy area next to the Green Building] for a barbecue and friendly conversation."

A year later, Caltech students snuck a parody of *The Tech* (with headlines including "Architects Deem Campus 'Unfortunate'" and "Infinite Corridor Not Actually Infinite") into distribution at MIT. And on Thanksgiving weekend in 2009, they tried another full-fledged hack, with banners, a website, and *another* parody of *The Tech*—all declaring that Caltech had bought MIT and would turn it into the "Caltech East School of Humanities." Delayed by MIT's almost-around-the-clock janitorial work, though, they were caught raising a banner in the front of Building 7 and had to take down all their signs.

"The goal of this prank was to play at the friendly rivalry between our schools, and at the same time do something awesome, not to be obnoxious or taunting," one Caltech East prankster said. "I don't know when Caltech will try something on this scale again, or when MIT will try a response, but I can't wait to see what's tried next!"

THE DOMESNIGHT BOOK

"I'll probably be here for awhile; I understand a bunch of engineering students put my motorcade on top of Building 10," President Barack Obama joked during a speech on clean energy at MIT in October 2009. He referenced, of course, the string of famous hacks that have appeared on the Great Dome of Building 10, which rises 150 feet above the campus. In the past decade, this tradition has prospered, but it also reflects a more general trend in hacks.

"We see many more hacks that are themed around anniversaries and theatrical releases, and fewer that address specific MIT issues," commented André DeHon (1991 SB, 1993 SM, 1996 PhD), a member of the advisory board for the IHTFP Hack Gallery, a web repository for MIT hacks: <http://hacks.mit.edu>. "I wonder what that means about the evolution of the culture," he added. "Either students are happier and have fewer things to protest. Or else it's about the world being more connected, although it's not clear to what extent people are playing to a global audience and a global audience is paying attention."

One of the most spectacular event hacks was the fire truck that appeared on the Dome on September 11, 2006, the fifth anniversary of 9/11, and was thought to honor all rescue workers. The truck was 25 feet long, about as big as the top of the Dome can accommodate, festooned with Dalmatian dogs and a fire hose. An inflatable fireman sat inside the cab, whose roof held a rotating red light that flashed at night.

"It was monumental hack," recalls Eric Schmiedl (2010 SB), who was allowed up to the top of the Dome to take the photos shown here (and who has taken many of the other gorgeous hack photos in this book). "It looked great even up close, which is a lot more than you can say for many hacks."

Another popular favorite is the replica of the Wright Flyer, plopped on top of the Dome on December 17, 2003—the one hundredth birthday of the first controlled heavier-than-air powered flight. "Given the gray December weather, it likely encountered turbulence while coming in for its landing," the IHTFP Hack Gallery notes. While the aircraft did not survive disassembly, the MIT Museum does have the vehicle's airworthiness certificate.

In 2009 came another all-time favorite hack that seemed to refer only to the MBTA's Red Line subway, which is not widely celebrated in song and story. On the morning of April 27, a two-car train zipped back and forth on the Dome, powered by the sun and cheered by a large crowd.

On the fifth anniversary of 9/11, hackers paid unusual tribute to the heroic firefighters of the NYFD. The logo painted on the truck, *Meninimus*, translates to "We remember," 2006.

The Apollo 11 landing module hack was still up on the Dome during the first day of final exams, when this 20-foot statue of Athena appeared in Killian Court, 2009.

Solar-powered MBTA Red Line cars on Dome, 2009.

THE THEATRE OF HACK

Off the Dome, many hacks have honored movies and games.

When the movie adaptation of *The Hitchhiker's Guide to the Galaxy* series came out in 2005, for instance, the radar on top of Building 54 became a good likeness of the iconic image of the planet sticking out its tongue that was used on the cover of the original books. A related banner hung off Building 10 with the books' second-best advice: "Don't panic." (The first, "a towel is about the most massively useful thing an interstellar hitchhiker can have," was commemorated in 2001 by a towel hung on Building 10 along with a banner honoring the passing of *Hitchhiker* author Douglas Adams, who had given a reading of a new book the previous year on campus.)

A brilliant modification of the radome on top of the Green Building marking the release of *The Hitchhiker's Guide to the Galaxy* movie, 2005.

Harry Potter clearly has fans in the hacking community. When the sixth book in the series was published in July 2005, a replica of the hero's scar cropped up on the Dome. In November of that year, when the fifth movie debuted, signs turned Building 9 into Building 9 3/4 (Track 9 3/4 being where wizard children take the train to their boarding school) and changed its signage accordingly (bathrooms were labeled "witches" and "wizards," for instance). On the day the final book in the series arrived in July 2007, a Dark Mark (symbol of evil Lord Voldemort) glowered over the Student Center and the Ray and Maria Stata Center offered a place to park your broomstick.

On the official day of publication for the final book in the Harry Potter series (*Harry Potter and the Deathly Hallows*) MIT students awoke to find Lord Voldemort's "Dark Mark" on the Student Center, 2007.

In the *Halo* videogames, the main protagonist, Master Chief, is named John. As the highly awaited *Halo 3* arrived in September 2007, MIT hackers realized that the character was based on John Harvard, and they suitably transformed the famous statue on that other Cambridge campus. At dawn, John Harvard wore the game's helmet, carried the game's assault rifle, and had a Beaver emblem on one shoulder.

The graphic novel and film *V for Vendetta* feature another totalitarian-fighting hero: V wears a mask of Guy Fawkes, who attempted unsuccessfully to blow up the British Parliament building on November 5, 1605, an event still celebrated each year in England. On that morning in 2007, *V for Vendetta* masks cropped up on items across campus. In the evening, students wearing the masks and black robes marched across campus to an Undergrad Association Senate meeting in the Student Center. After their leader delivered a short speech, the group fired party poppers into the air and walked out.

John Harvard revealed as *Halo 3* hero by MIT hackers the night prior to the video game's release, 2007.

BIG MESSAGE ON CAMPUS

As administrations officer Ben Jones once remarked, "hacking is the students reminding the administration that they are smarter than them." The stream of hacks that protest, comment on, or simply spoof Institute events may have dropped but has not dried up.

Many of these have targeted the buildings completed in recent years, several of which have what might be called distinctive designs. Among these, the Ray and Maria Stata Center (Building 32) stands out with its playful, irregular, and frequently disparaged architecture.

The morning that the Stata Center was dedicated in May 2004, one high wall was labeled with a mammoth MIT Property Office sticker complete with a barcode that lacked any right angles, just like the building itself. In 2006, hackers added a matching "V" and "O" to the stainless steel "MIT" sign in front of the Stata Center, "perhaps to express their opinion of the architecture, or perhaps as a commentary on the start of term," the IHTFP Hack Gallery notes. In December of that year, you could see a Christmas tree with suitably strange angles in a Stata lobby, and one Stata tower added glasses and a red-and-white cap to impersonate the star of the *Where's Waldo* children's books.

Hackers could not resist adding a couple of letters to the shiny stainless steel MIT sign outside the Ray and Maria Stata Center, 2006.

Stata-as-backdrop continued during Campus Preview Weekend in 2008, when a UFO appeared, complete with a stuffed alien escaping through the smashed windshield and a "hole" through the wall in back behind it. Also that spring, the rash of amusing "Do Not" stickers dabbled around the campus included several on Stata, one being "Do Not Straighten This Building."

Other hacks, of course, took on Institute twists and turns. One classic came in 2001 with a funeral ceremony during registration for Rush, an event which for more than thirty years had allowed incoming freshmen to pick where they would live (see "Beyond Recognition: Commemoration Hacks").

On registration day two years later, a group of "Special Police of the Office of the Dean for Student Affairs," clad in black and wearing sunglasses, set up booths to hand freshmen bogus security forms. The forms then were fed into a security-enhanced fax machine (also known as a shredder). The freshmen themselves were stamped as Registered.

After the Ocean Engineering department was merged into the Mechanical Engineering department in 2005, a pirate hung from a mast standing in the snow outside the Kresge Auditorium, with two tombstones for the department.

That year, the famed and often-hacked mural of a dollar bill that stood for many years outside the cashier's office in Building 10 suddenly reflected the inauguration of President Susan Hockfield. George Washington's quiet stare was replaced briefly by a smiling portrait of the new president. (The following year, just before the mural was taken down because the cashier's office was being replaced by a student lounge, George's giant image shed a tear.)

The venerable string of MIT commencement hacks stretched farther in 2007. As Chair of the Corporation, MIT's governing body, Dana Mead took the podium, five panels unfurled between the columns of Building 10, saying "I'll Have Thesis Finished Pronto." The phrase's acronym, "IHTFP," is a student slogan at MIT for their feelings toward the Institute and is incorporated into many hacks (see "Intriguing Hacks to Fascinate People").

"Presumably as a commentary on the state of the MIT student being chased by the monster of final exams, on the morning of December 9, 2008, hackers turned the Mechanical Engineering Lounge in Building 3 into an 'aquarium' featuring a sea-monster with scales made of old final exams, a sandcastle, Tim the Beaver in a scuba outfit swimming away from the sea-monster, seashells, and bubbles in the windows," noted Eric Schmiedl.

On the day of Susan Hockfield's inauguration as MIT's sixteenth president, hackers substituted her portait on the famous dollar bill mural outside the former MIT Cashier's Office, 2005.

ONE OF A KIND OF A HACK

Once again, the past decade brought hacks that have the common thread of not much in common, aside from the joy and humor of idea and execution.

A prime example is the 2003 Robot Rights Protest, held outside the annual 6.270 robot games competition to rally against robot slavery and mistreatment. Celebrity headliners included R2D2 from the *Star Wars* movies and Hal 9000 from *2001: A Space Odyssey.* The protesters' handouts included a plea to ban the Second Rule of Robotics, which forbids any robot to disobey the orders of any human unless it brings harm to other humans. "While slavery is banned for humans, robots must face legalized slavery everywhere," the protestors declared. "Contemporary slavery takes various forms and affects bots of all ages, sex, operating systems, and machining."

Just prior to the start of the Electrical Engineering and Computer Science 6.270 course competition, robots stage a protest, 2003.

Nonhumans also headlined another 2003 protest, as hundreds of gnomes invaded the Athena computing center in the Student Center. The gnomes came in every variety, plaster or plastic. At least one had a curious resemblance to a prominent denizen of the computing center. Posters for Acme Gnome Exterminators were pinned up nearby.

A bit late, the following year, quietly bogus No Trespassing signs began cropping up around campus. The small black stickers carried rather special messages, such as "Access to and presence in this building is limited to rodents and other pests. Venom will be produced.") "The stickers looked so genuine that they could still be found in various places around MIT months after the hack went up," according to the IHTFP Hack Gallery. Another 2004 hack lasted only minutes and was meant primarily for just one viewer, Elizabeth Ann (Lizzie) Hager. One day during finals week in May 2004, she found herself in Lobby 7 along with her boyfriend, Ed Barger, and Ed's family and a rather large number of friends. Ed pointed out a banner partially unfurled from the roof that said: "I love you." Cool hack, Lizzie thought. Then the sign kept unfurling to say, "Lizzie, will you marry me?" Ed got down on one knee and presented a ring, and Lizzie accepted. They married the following year.

On a less romantic note, MIT, like most academic centers, is filled with programmers who may prefer the open-source operating system Linux to Microsoft Windows. That stance has produced a few awkward moments during visits by Microsoft founder Bill Gates, who has contributed generously to MIT. In fact, on the morning that the building carrying his name was dedicated in May 2004, the Internet kiosks in the lobby suddenly were running Linux rather than Windows XP and showing a welcome from Tux the penguin, the Linux mascot.

Perhaps the strangest Linux tribute—if that's what it was—came in April 2006 when a "Peace, Love, and Linux/Penguins" logo created entirely in multicolored condoms could be viewed in the skylight of a Building 32 lecture hall. (The logo came from a marketing campaign backed by IBM.) The lecture hall also was papered with Linux flyers declaring "Practice Safe Hex—Avoid Computer Viruses Today!" and scattered with more condoms.

In a less esoteric and far more widely seen exploit, hackers sprinkled almost a hundred yellow cranks, modeled on those that power the Media Lab's XO laptop (from the One Laptop Per Child project aimed at building and distributing inexpensive computers to children in developing countries) around MIT in February 2007. The Wiesner Building, home of the Media Lab, itself received a large yellow crank. Other cranks materialized on everything from computers to a restroom door.

MIT programmers love open-source software. Hackers urged students to practice "safe hex" with flyers and this unusual rendition of the Linux marketing campaign graphic made from condoms, 2006.

Modeled on a One Laptop Per Child program's XO design, the yellow cranks popping up across campus in 2007 had already boosted the campus's energy efficiency by 0.0005%, or so their accompanying posters boasted.

One morning in April 2007, the three round windows in Building E25 that face the courtyard suddenly became a traffic light, with a Walk button on a nearby post. The red light lit up as "Hack," the yellow light said "Punt" (to skip certain work because of competing demands, in MIT lingo), and the green light was marked "Tool" (to grind away at work, or someone who does). The installation briefly became a magnet for games of Red Light/Green Light.

The three round windows on Building E25 have tempted hackers before, but this was the building's first-ever working traffic light, 2007.

Games went big-time that December, when the campus went wild with board-game hacks, cleverly leveraging the idiosyncrasies of MIT architecture.

Campus maps became Risk boards, with plastic soldiers attached. The Stata Center's Student Street morphed into Mouse Trap, with the Hilltop Steakhouse cow (part of the MIT Museum exhibit on hacks) as the mouse. A game of Scrabble played on the Media Lab's gridded exterior,

A large version of Settlers of Catan was one of a half-dozen board games that appeared around campus in 2007.

featuring words such as "Punt" and "Athena." Students challenged administration in chess on the paving stones in the courtyard behind the Media Lab. Cranium dice popped up in the lobby of Building 34, with Cranium game cards played in the lobby of Building 46. And a large Settlers of Catan board surfaced on the Dot, "complete with an MIT Campus Police officer as the 'robber' character standing watch with a can of A&W (root) beer in his pocket and a doughnut in his hand," noted the IHTFP Hack Gallery. "The board featured a 'Building Costs' card that listed an MIT education with a cost of 160,000 cash and a return of 10 points."

MIT's fascination with things zombie, which long predates the current wave of zombie movies and literature, also had its day. "As we all know, zombies feast on brains, juicy brains," student Charles Lin commented. "And unfortunately, the brains at MIT are pretty juicy and delicious." That may be why zombie marches are staged not infrequently at the Institute. Perhaps worried about the impression on prefrosh, hackers responded accordingly for the 2008 Campus Preview Weekend, putting "In Case of Zombie Attack, Break Glass" emergency boxes with chainsaws in Buildings 16 and 46. Of course, as one student soon remarked, "everyone knows zombies can break glass."

Concerns about body parts also enlivened the Stepford Labs hack in February 2010. Spoofing the new major in biological engineering, hackers "installed a display case full of 'enhanced' simulated body parts in MIT's Infinite Corridor," Eric Schmiedl said. "The body parts included: a head with a functional video camera replacing an eye, a leg with a power socket, feet with rolling wheels, an *Avatar*-style head, a face with LEDs in the eyes (transmitting 'IHTFP' in Morse code), a head with a *Matrix*-style socket in the neck, a neck featuring a jack for "IP over Voice," as well as analog audio and a hand with a USB "thumb drive."

Best of all, many viewers of the Stepford Labs display didn't even realize immediately that it *was* a hack.

COMMEMORATED, COURTED, CAUGHT, OR CONDEMNED?

Hacking hasn't been just fun and games in the past decade. Less happily, we've also seen rising though infrequent friction between the hacking community and the administration, which generally remains accepting of hacks but whose uneasiness about their fuzzy legality and their safety risks has risen sharply since 9/11.

Because of legal and insurance liabilities, the administration cannot encourage students to go prancing around rooftops at night, particularly after an undergraduate was injured by falling through a skylight in 2006. But such serious and potentially tragic mishaps typically reflect an

Many MIT labs now have corridor displays of their work. Hackers so cleverly developed this mock display of the Stepford Labs that some viewers did not recognize it as a hack at first, 2010.

individual student's state of mind rather than hacking, according to David Barber, who is the MIT emergency and business continuity planner, a longtime hack remover, and an unofficial liaison between the administration and hackers. "The hacking community always, always has safety as the number one issue," he emphasized. "The last thing they want is for anybody to get hurt. They're not out for their personal glory."

Other administration concerns have grown from a small series of episodes that blurred the uncomfortable line between hacking and breaking and entering, at least in the eyes of the campus security force. "I'd rather deal with another prison riot than this hacking," said MIT Director of Facilities and Security John DiFava, a former state police officer, in a 2008 interview with *The Tech*.

An incident in which three students were arrested in the Faculty Club late one Saturday night in 2006 caused a particular ruckus after it became public in February 2007. The case was moving toward trial in Cambridge courts when it was dropped after negotiations between MIT and the district attorney's office. MIT faculty and students who felt that the case should have stayed within the Institute reportedly raised well over $10,000 for the students' legal costs.

This controversy helped to prompt an October 2007 letter from Chancellor Phillip Clay, written with much input from faculty and students. "The incidents that give us pause come with a concerning frequency," Clay stressed. "Hackers or want-to-be-hackers have suffered serious injury and narrowly escaped much worse in recent years. Other incidents have put students (and MIT) in awkward positions in relation to law enforcement agencies or brought notoriety to the Institute. This is unacceptable."

The chancellor repeated the message in another email sent in September 2008. "Hacking is the design and execution of harmless pranks, tricks, explorations, and creative inventions that demonstrate ingenuity and cleverness," Clay said. "Those who violate the tradition, by endangering themselves or others, by breaking the law or other departures from the 'hacking code of conduct' cannot seek protection from responsibility, and they will be held accountable for their actions."

Response was prompt.

Four days later, the historic hacks showcased in the MIT Museum display at the Stata Center were covered with black and hung with violation notices purporting to explain how they had broken the rules mentioned in Clay's email.

Nothing prompts a quick protest like being told how to hack by the MIT administration, 2008.

Those who felt the waters were muddied also could point to the daylong, entirely official fiftieth celebration of the Smooting of the Harvard Bridge the following weekend (see "Smoot Points").

While this controversy has simmered down, at least for the moment, many in the hacking community think that many in the administration fail to understand hacking's place in MIT culture. "The people who pull hacks aren't the ones who will roll over and say, 'Yes sir,'" said one former hacker. "I'm surprised that there aren't more hacks about this. If the Institute says they can't do X, they do X."

HANDING ON THE HACK

At the start of Campus Preview Weekend in 2010, a distinctive hack appeared high in the air on the *underside* of the Media Lab arch: An inverted lounge furnished with chairs, sleeping cat, whiskey bottle and glasses, potted palm, and pool table that was complete with cues, balls, and plans for a hack on the Great Dome. "Pretty awesome!" one visiting prefrosh commented.

Never mind administration strictures, the never-ending academic ultra-pressure, social pressure, and new interactive channels such as social media networks: All the old and new distractions for MIT students have not stopped the flow of hacking or its typical schedule, often targeting times or events where hacks will gather the biggest audience. Hacks are likely to appear around the beginning of term, on Halloween, in late December and the January Independent Activities Period, for April Fools' Day and Campus Preview Weekend, at the end of semester and graduation, and (for patriotic reasons) on the Fourth of July.

"The blockbuster hacks are crazy big engineering projects," Eric Schmiedl noted. "Smaller hacks and exploration hacks are much more common. A big hack is like a final project, taking all the knowledge and experience accumulated from smaller hacks. Hacks are MIT's unofficial school of management."

"I continue to believe that there's a huge value to hacking for students, from several sides," André DeHon said. "On one side, it does help in balancing life. Another side is what you get out of it as engineering practice. And there's also a philosophical value for having an incomplete acceptance of authority, in always questioning things whether they are scientific or not."

"Technology changes, and it changes rapidly," DeHon added. "The challenges and targets for hacks also are changing rapidly, and that's healthy. Good hackers are off thinking, what will people not expect? There's an artistry to that."

Campus Preview Weekend is a favorite time for hackers to strike. In 2010, they installed an upside-down lounge complete with chairs, lamp, pool table, and even a sleeping cat.

WHERE NO COW HAS GONE BEFORE: ACCESSING THE INACCESSIBLE

Armchair aficionados of the sport often assume that hacking began as a twentieth-century phenomenon. But even before the Institute crossed the Charles River from Boston to Cambridge in 1916, MIT students were hacking. John Ripley Freeman, renowned civil engineer and member of the class of 1875, noted in his memoirs that pranksters habitually sprinkled iodide of nitrogen, a mild contact explosive, on the drill room floor, adding considerable snap to routine assembly.

Of course, pranksters were not called hackers back then; only within the last forty years has the term "hack" been synonymous with campus hijinks. But it was in those formative years —well before the term "hacking" was coined—that the spirit and traditions of the sport at MIT were established.

Institute hacks in the late-nineteenth and early-twentieth centuries were primitive by today's standards—jokes played on professors or pranks sparked by interclass rivalries. Freshmen would steal the sophomore class flag before the annual football game. Sophomores would rearrange the furniture in freshmen dorm rooms while students were at class meetings. The early MIT yearbooks featured a section called "Grinds" filled with cartoons, doggerel verse, and descriptions of pranks.

In the 1920s, aficionados have noted a change in these student pranks. While early on these hijinks were mainly spontaneous, now some of these stunts and jokes seem to be a bit more organized. It was then that campus hijinks increasingly began to be attributed to the Dorm Goblin. The Dorm Goblin was not a specific group of students but rather a mythical entity similar to Kilroy of World War II fame. But the fact of attribution is one reason that many today consider these Dorm Goblin pranks of the 1920s the direct antecedent for the hacks unleashed on the Institute by organized student groups beginning in the 1960s.

The Dorm Goblin pranks included the usual college fare, such as the threading of a 35-foot telegraph pole through Senior House dormitory in January 1928 or coaxing a live cow to the roof of the Class of 1893 dormitory (now East Campus dorm) dorm a few months later. The cow went up fairly happily but was none too pleased to make the trip down, teaching MIT students a basic fact every college student involved with "cow pranks" seems to learn the hard way.

Beyond cows, the Dorm Goblin became associated with more technical pranks, like turning dormitory phones into radio speakers, which allowed students to listen to the latest tunes in their rooms by taking the receiver off the hook. The Goblin was also more than likely responsible for launching the door hacking tradition of disappearing, changing, or redecoration of doors that persists to the present.

Live cow on top of the Class of 1893 dormitory, 1928.

DORM ROOM WITHOUT WALLS

A full dormitory room on the Charles River, 1985.

With the advent of the Dorm Goblin, MIT students began to take greater notice of the various pranks. Accounts of the latest Dorm Goblin stunts began to be published in the student newspaper, *The Tech*. The most popular pranks were those that seemed to conquer the inaccessible and make possible the improbable. In 1926, for example, the Dorm Goblin was credited with bewitching a telephone booth. *The Tech* reported: "After a man had been telephoning for a time longer than the Dorm Goblin thought reasonable, the booth was bewitched and performed some amusing antics. [The booth] tried to do a mild imitation of the Charleston, tipped part way over, and so surprised was the occupant that he put his fist through a quarter-inch plate of glass in the door." Today, the avoidance of injury is one of the key requirements of a hack but a dancing phone booth, (would that we still had such things!) would clearly still be considered an interesting hack even now.

In his autobiography, *Surely You're Joking, Mr. Feynman!,* Richard Feynman (1939 SB) recounted his experiences with door hacking. A Nobel Prize–winning physicist, Feynman was equally famous for his practical jokes. What amused Feynman most about his door hack was that although he acknowledged his role almost immediately, he had already earned such a reputation as a wise guy that the other students thought he was kidding and did not believe him.

RICHARD PHILLIPS FEYNMAN
ΦBΔ

Far Rockaway, New York

Born May 11, 1918; Prepared at Far Rockaway High; General Physics; Dean's List 6; Physical Society (2, 3), Vice President (4); Entered Freshman Year.

Richard Feynman, *Technique 1939*.

WHY WE HACK
Anonymous

Many of us hack first and foremost because it is fun. It's fun to scale tall buildings and plant unusual objects there. It's fun to do difficult things and do them well. It's fun to make people smile, and it's fun to make them scratch their heads in wonder.

Most hacks represent a feat of engineering—sometimes it is electronics or mechanics, and sometimes it's the logistics of getting six people and an eight-foot-tall papier-mâché beaver through the halls of the Institute at 2 a.m. without arousing the suspicion of the campus police or custodial staff.

Other hacks are well-timed art, where the beauty is in the elegance of the medium and the appropriateness of the message, coupled with the challenge of deploying the art visibly without official sanction. One thing common to almost every hack is that it involves doing something most people wouldn't think of doing. It should come as no surprise that hacks abound at MIT, a place dedicated to doing things other people haven't thought of doing.

How about anonymity? Secrecy helps us do our job, certainly, but there is also a modesty that infuses the engineering tradition. The artist should not get in the way of the art, and the designer should not get in the way of the design. No amount of advertising or cajoling can turn a bad design into a good one; instead you put your work out for public scrutiny and let others evaluate. Anonymity is not without its perks. It's quite a thrill to have someone point out your hack to you as something you must see.

A successful hack brings the satisfaction of having brightened the days of many people. An unsuccessful hack teaches valuable principles of engineering—plan ahead and check theory with experiment. What better pastime for aspiring scientists and engineers?

"Portrait" of hackers just before completing their work, 1987.

Tech Dinghy floats in Alumni Pool, 1987.

AUTO INDUSTRY: THE GREAT VEHICLE HACKS

As the automobile grew increasingly central in American culture, it became a featured element of many hacks. During the 1920s and '30s, MIT pranksters hauled cars up the sides of dormitories, parked them on the front steps of buildings, and hid them in basements—they even impaled a Ford coupe on a steel pole.

In January 1926 the Dorm Goblin surreptitiously, and with nary a scratch to the vehicle, moved an illegally parked touring car to the basement of the Class of 1893 dormitory (now East Campus dorm). The removal required a team of sixteen workers and a tractor. Then in March, the Goblin (a team of fifty students) struck again, this time hauling a Ford chassis, with engine intact, up five stories to the roof of the same dormitory and perched the front wheels rakishly over the side. According to myth, at least one of the photos of the event pictures James Killian (member of the Class of 1926, Institute president from 1949 to 1959, and Killian Court namesake) as the triumphant driver.

Ford chassis on the Class of 1893 dormitory, 1926.

Students pull a car, which reads "Tech 2 Hell," up the side of the Class of 1893 dormitory, 1936.

This was a popular prank, as students pulled another car (bearing a popular MIT student complaint: "Tech 2 Hell") up the side of the same dorm in 1936, and hackers parked a car on the top of the stairs to Walker Memorial, the MIT student union building, in 1959. In 1985, hackers made an artful placement of a Volkswagen Beetle at the entrance of the Zeta Beta Tau fraternity. The Bug's nose jutted out over the front steps as though it had just burst out of the frat house following one of those wild rides depicted in cartoons.

Rooftop view of students using painter's block and tackle to bring the car up the side of the Class of 1893 dormitory, 1936.

One of the most legendary automobile hacks was also perpetrated in 1985: The Massachusetts Toolpike ("tool" is an MIT expression for studying, and the Massachusetts Turnpike is a major interstate heading west out of Boston) encompassed the entire central hallway of the Institute. Hackers divided the 775-foot Infinite Corridor into lanes with yellow tape, added a rotary at the intersection of two corridors, and even created angled parking spaces, with a neatly parked real car in one. The sign at the "entrance" said: "Massachusetts Toolpike: Toll $16,000" (a year's tuition at the time), and posted highway signs instructed travelers about regulations and tolls, such as "SPEED LIMIT 3×10^{10} cgs." At the Toolpike's terminus at 77 Massachusetts Avenue, the final sign read: "Exit 77 Real World." This vehicle hack remained unsurpassed for nearly a decade, until one of the finest hacks of all time: Campus Police Car on the Dome. But hacks on the dome fall into a category all to themselves.

The speed limit for the Massachusetts Toolpike, 1985.

The toll advisory for the Massachusetts Toolpike reflected the cost of tuition, 1985.

A car parked in lobby 10 as a part of the Massachusetts Toolpike, 1985.

DOMEWORK: HACKING THE DOMES

The Institute's two signature domes have always been popular venues for hackers' surrealist dioramas. The Little Dome is 72 feet in diameter and 100 feet high, and the Great Dome is 108 feet in diameter and 150 feet high. At the top of each is a flat platform approximately 20 feet in diameter (the Great Dome's is slightly larger). The appearance of a cow, a prosthetic device, or a dorm room is all the more dramatic because access to the domes is difficult and relatively closely monitored. After making it through the 3-by-4-foot door, a hacker still has to scramble up a ladder (self-supplied) before ascending the curved slope of the dome. A good deal of the art of a dome hack is the elegance of the solution hackers find to overcome these challenges.

In the predawn hours of May 9, 1994, early-rising students were drawn to lights flashing on the dome. As the sky lightened, the whir of traffic helicopters alerted other students and staff. Campus Police Chief Anne Glavin heard about the situation on the news while driving to work. Gradually, all could see what has since become one of most famous and best-loved hacks: Campus Police Car on the Dome.

The luminous spring sky proved an ideal backdrop against which to view the white car glinting in the sun 150 feet above the ground on the grand dais of the Great Dome. On the ground, spectators could see the flashing bar of emergency lights across the police car's roof. The helicopter traffic reporters could make out the words "MIT Police," but it was not until later that the amusing details of the faux cruiser would be known. As with most hacks, it was both more and less than it appeared to be.

The car was a bit of a stage set. Hackers had attached segments of the outer metal shell of a Chevrolet Cavalier to sections of wooden framing making it possible for them to fit each piece through the small door that opens onto the area at the base of the dome. Once on the dome, the hackers had reassembled the car by bolting the frame together. They installed a mannequin dressed as a campus police officer with a toy gun, a cup of coffee, and even a box of donuts (still preserved in the MIT Museum collection!). A pair of fuzzy dice hung from the rear-view mirror. The car's number was π, the vanity plate read "IHTFP," and a small plastic yellow sign in the rear window warned "I Break for Donuts." The hackers' final touch was a parking ticket affixed to the front window. The alleged violation? "No permit for this location"!

By 10 on the morning of its unveiling, MIT facilities workers had removed the hack, but news of the event proliferated. The story hit the evening news and the wire services spread it

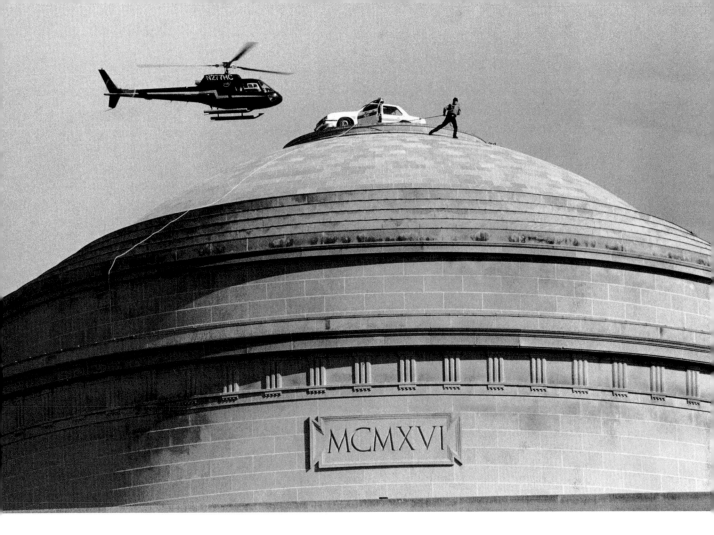

Campus police car on the Great Dome, 1994.

Hackers from the Burton One Outdoor Breast Society show some of the challenges associated with putting a hack on a dome.

worldwide. At the campus movie that evening, the ads before the previews included this appeal: "Missing: One white and blue patrol car. If found, call x3-1212" (the number for campus police).

DOME AS DAIS

The very first class picture (Class of 1920, entering in the fall of 1916), taken in front of MIT's new Cambridge campus, shows students on the ledges and rooftops of the buildings known as the "Main Group." No one knows for sure when the very first Dome hack occurred. As MIT celebrates its 150th anniversary, however, it should be noted that on April 10, 1961, members of the freshman class celebrated the Institute's centennial with a 9-foot-tall cardboard candle atop the Great Dome along with a happy birthday banner. Over the years, there have been more than fifty dome hacks. Some of the favorites include:

GREAT PUMPKIN, 1962 AND 1994

On Halloween night in 1962, hackers transformed the Great Dome into a smiling jack-o'-lantern. An officer from the fraternity believed to be involved with the hack explained in *The Tech* that the hack had been inspired by the cartoon character Linus from *Peanuts* by Charles Schultz. "It seemed a humanitarian thing to do—to make the 'Great Pumpkin' rise out of the pumpkin patch on Hallowe'en so that the Linus's of the world would not be disillusioned." Hackers continued the tradition by mounting a second Great Pumpkin on the Dome in 1994.

A tribute to the Great Pumpkin, 1962.

Sometimes student hacks have inspired imitations. Such is thought to be the case when a construction crew working on the renovation of the MIT Press Bookstore building in 1992 created its own jack-o'-lantern on the orange mesh surround the building. The idea probably originated with MIT staff who told the crew about student hacks, but it is not known if the 1992 stunt was an homage to the original Great Pumpkin hack.

Halloween decorations appeared during renovation of E28, 1992.

KILROY (AKA GEORGE) WAS HERE, 1972

In an amazing feat of "textile engineering," hackers celebrated Halloween by recreating the World War II icon Kilroy with 6,000 square feet of polyethylene sheeting. An "interested resident," E. Martin Davidoff, wrote a letter to *The Tech* bemoaning the quick removal of "Kilroy" (or "George"): "What harm would there have been to let poor George exist for one full day? The spirit of the Institute dies again as the bureaucracy acted efficiently only in destroying something that only could've brought a smile and chuckle to the people who normally trudge in and out of the 77 Massachusetts Ave. entrance." A week later, George reappeared and was allowed to remain for a full day.

Kilroy (aka George) on Building 7, 1972.

THE BIG SCREW, 1977 AND 1985

The screw is a recurring motif in the social history of the Institute. Every spring, MIT's chapter of the national service fraternity Alpha Phi Omega (APO) runs the Institute Screw Contest. The Big Screw is awarded to a faculty or staff member deemed to have been the most successful at "screwing" MIT students. The selection is based on votes, or rather the total amount of cash donations made in the nominee's name. (The winner of the contest gets to designate the charity that will receive all the funds collected.) Given the popularity of the contest, it is not surprising that over the years both domes have been "screwed."

Left-threaded wood screw in the Great Dome, 1985.

THE GREAT BREAST OF KNOWLEDGE, 1979

A hacking group by the name of the Burton One Outdoor Breast Society finally accomplished its objective to adorn the dome with the nipple that has been conspicuous by its absence for three quarters of a century.

THE GREAT BREAST OF KNOWLEDGE

Brian Leibowitz

In his 1990 Journal of the Institute for Hacks, TomFoolery & Pranks at MIT, *Brian Leibowitz gives a behind-the-scenes account of the design and creation of the "The Great Breast of Knowledge" hack—a case study in the skills and perseverance necessary to execute a hack on the Institute's Great Dome. Here is an excerpt.*

The notion that the Great Dome resembles a giant breast was first suggested in the living group Burton One in December 1978. During MIT's Independent Activities Period (IAP) the following month, blueprints of the dome were studied and the proper dimensions for the nipple and areola were determined. The nipple was then constructed using a wooden frame covered with chicken wire and pink paper.

As hackers prepared to haul it up the side of the building, campus police came upon the nipple sitting on the roof of the hackers' car. Undaunted by this setback, the group planned their second attempt. This time, practice sessions were held to hone climbing skills and to reduce the time needed to transfer the nipple from the car to the roof. Alas, preparations notwithstanding, implementation was foiled again—the hackers were spotted by a cleaning woman and decided to postpone the hack.

Alleged treachery barred the Burton group's third attempt, when campus police, acting on an anonymous tip, arrived at the scene to stop what they had been led to believe was a robbery in progress. Their escape hampered by a snagged rope, two members of the hacker roof crew were caught. After a chat with the campus police chief, the students agreed not to try again that semester. Later learning that the anonymous tipster may have been a student from another floor of their own dorm, the hackers realized that secrecy was now essential to their success.

A new plan was formed around the annual freshman picnic that marked the beginning of a new semester. The redesigned, collapsible nipple would not have to be hoisted up the side of the building but could be carried in backpacks instead. On the eve of this fourth attempt, the Burton hackers announced to their dormitory that the project had been called off. Finally, during the picnic, the Burton One Outdoor Breast Society successfully installed the Great Breast of Knowledge with its accompanying banner, "Mamma Maxima Scientiae."

The Burton One Outdoor Breast Society's nipple on Great Dome, 1979.

STEER ON THE DOME, 1979

On Halloween, hackers "rescued" a life-sized fiberglass steer from its grassy knoll in front of the Hilltop Steakhouse and gave it a more prominent grazing field on the Great Dome. When "Ferdi" was returned, the Hilltop management placed a mortar board on its head and a diploma in its mouth. When the MIT Museum opened its Hall of Hacks exhibit in 1991, the restaurant donated Ferdi to MIT. The steer is now perched above the Forbes Family Café in the Ray and Maria Stata Building, part of a permanent campus display of hacks maintained by the museum.

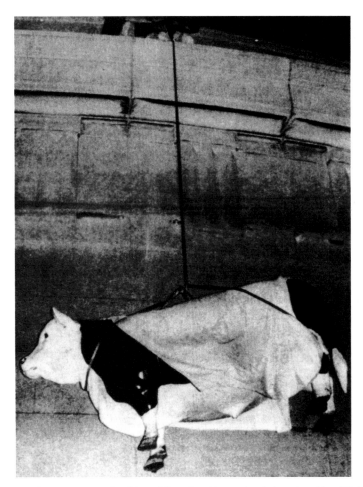

The fiberglass steer Ferdi from Hilltop Steakhouse hauled up the side of Building 10, 1979.

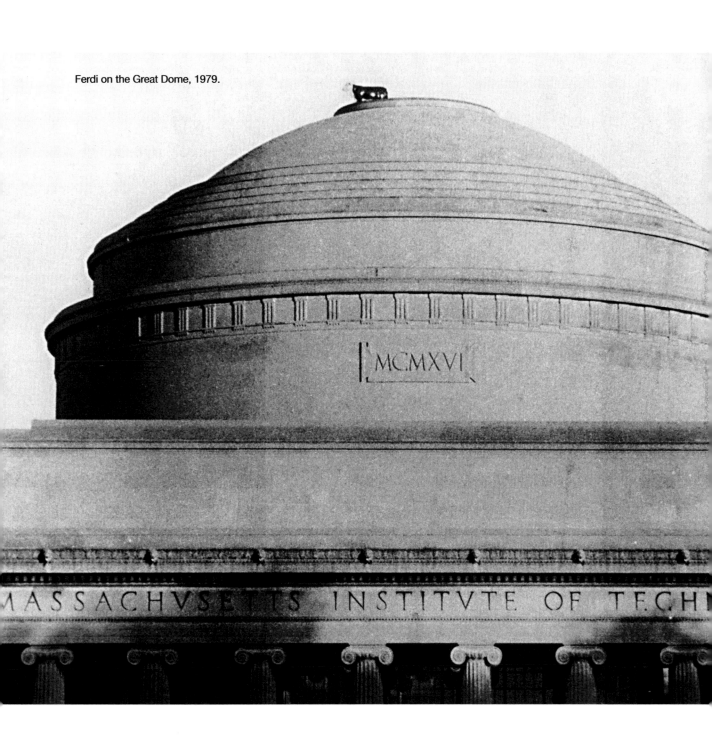

Ferdi on the Great Dome, 1979.

PHONE ON THE DOME, 1982

When a campus policeman scaled the dome to investigate, he discovered that the telephone booth was indeed the real thing. In fact, the booth light was on and the phone was ringing. When he asked for advice from campus police headquarters about what to do, you can almost hear the laughter in the reply: "Well, answer it!"

HOME ON THE DOME, 1986

To alleviate the problem of overcrowded dorms, the Technology Hackers Association constructed a 12-foot-high, 16-foot-square house, complete with mailbox and welcome mat. The twenty-eight panels of Room 10–1000 were hauled up the side of the building and secured with ropes and cables.

SNOWMAN, 1987 AND 2001

On a steamy August day in 1987, hackers erected a papier-mâché snowman on the Little Dome. Fourteen years later, in March 2001 they made one from real snow.

HOLIDAY LIGHTS, 1990 AND 1993

Hackers spelled out "MIT" with Christmas lights and placed several giant gift-wrapped presents on the Little Dome in 1990. In 1993, they illuminated the same dome with an 8-by-10-foot menorah for Chanukah.

WITCH'S HAT, 1993

For Halloween, hackers constructed what was purported to be the world's largest witch's hat on the Little Dome. Approximately 20 feet tall and supported by a 15-foot-wide brim, the hat weathered a very windy weekend before its removal on Monday morning—a tribute to the engineering skills of the hacking team.

CONTENTS UNDER PRESSURE, 1993 AND 2001

In 1993, hackers transformed the Little Dome into a giant pressure cooker with a massive dial and a banner that read "Warning: Contents Vnder Pressvre!" In 2001, another symbol of stress appeared during finals—this time, on the Great Dome. To the naked eye, the MIT landmark looked to have cracked under the weight of an 8-foot-tall 48-unit Acme weight—48 being the number of units in the standard 4–5 course load at MIT. Carefully secured with cables, the Acme weight was accompanied by a long strip of fabric mimicking a crack.

BEANIE CAP, 1996

It must have been the lunar eclipse the night before that inspired hackers to transform the Great Dome into a fully functioning propeller beanie. Red stripes alternating against the gray stone of the dome formed the cap, but the spinning propeller blades made this one of the most ambitious hacks. The hack even extended to the emergency vehicle used by MIT's Confined Space Rescue Team, which, among less amusing tasks, is responsible for removing hacks around campus. Called out to take down the Beanie Cap hack on the morning of September 27, the team found their truck relabeled the "MIT Hack Removal Team" and festooned with the images of five great hacks that the MIT facilities team had been responsible for removing in the past. The detailing included an "I brake for hacks" bumper sticker, a Hackbuster logo and the words "HACK REMOVAL" spelled backward on the hood of the truck. (See "It's Not a Job, It's an Adventure.")

CHEESEHEAD, 1997

When the Green Bay Packers, aka the Cheeseheads, beat the New England Patriots in the Super Bowl, hackers placed a giant chunk of faux cheese on the Great Dome.

GREAT DROID, 1999

Two days before the much-awaited *Phantom Menace* installment of the Star Wars movie series, hackers turned the Great Dome into the Great Droid. The beloved robot R2-D2 was created using fabric panels and a painted tent to represent the droid's holographic projector. Hackers thoughtfully included detailed disassembly instructions addressed to the "Imperial Drones" and signed "Rebel Scum."

THE EAGLE HAS LANDED, AGAIN, 1999 AND 2009

When Apollo 11 landed on the moon in 1969, its guidance computer was almost entirely an MIT product, and MIT alumnus Buzz Aldrin (1963 ScD) was the second person to walk on the moon. On the thirtieth anniversary of the first lunar landing, hackers created a man-on-the-moon diorama on the Little Dome.

In 2009, in anticipation of the fortieth anniversary of Apollo 11 and of a major MIT symposium sponsored by the AeroAstro department, a full-scale replica of the Lunar Module appeared on the Great Dome. Facilities gave it to the department for safekeeping, but hackers stole back their creation. Participants in the Giant Leaps Symposium (including Aldrin and fellow astronaut Neil Armstrong) soon discovered the Lunar Module had landed again, this time right outside Kresge, where it became a popular backdrop for many photos.

A working telephone booth on the Great Dome, 1982.

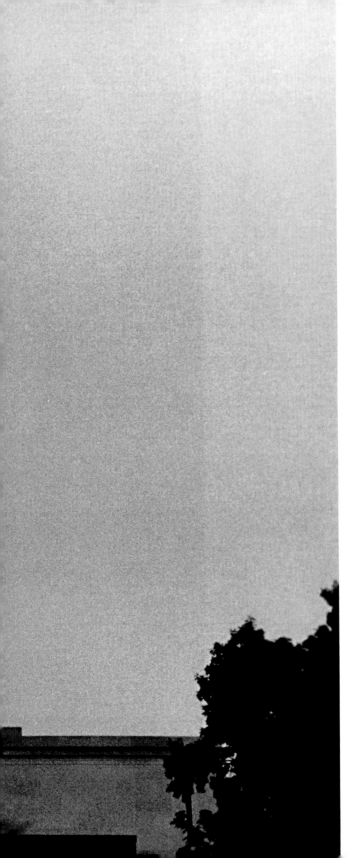

Home on the Dome on Building 10, 1986.

A papier mâché snowman on the Little Dome, Building 7, 1987.

Christmas lights decorate the Little Dome, Building 7, 1990.

An 8-foot-tall menorah on the Little Dome, Building 7, 1993.

Building 10 transformed into a pressure cooker, 1993.

A beanie cap adorns the top of Building 10, 1996.

Star Wars character R2D2 on Building 10, 1999.

The "Eagle" has landed; LM model appears on Building 10, 2009.

RING AROUND THE DOME, 2001

In anticipation of the release of the first Lord of the Rings movie, a giant replica of a famous gold ring appeared around the Great Dome. The One Ring was authentically inscribed with Tolkien's text, about which Gandalf remarked, "The letters are Elvish, of an ancient mode, but the language is that of Mordor, which I will not utter here." The official MIT hacking website, the IHTFP gallery (<http://hacks.mit.edu>), noted that hackers were "not completely successful, for neither the Great Dome nor Building 10 vanished into the realm of the shadows."

SPACEMAN SPIFF, 2004

Fans of the cartoon strip *Calvin and Hobbes* were saddened to learn that Spaceman Spiff (Calvin's alter ego) crashed his spaceship into the Little Dome on the first day of classes for the spring semester. A note from the hackers read "I'm really sorry I crashed my spaceship into the dome. The Zorg ambush known as the first day of classes was on me."

The One Ring from *The Lord of the Rings* appears around the Great Dome, 2001.

INTRIGUING HACKS TO FASCINATE PEOPLE

The origins of the acronym "IHTFP" are strictly anecdotal. Many have claimed the amorphous motto as their own. Its use has been unofficially documented in both the U.S. Air Force and at MIT as far back as the 1950s. Whatever its ancestry, generations of MIT students have delighted in the acronym's infinite versatility. "IHTFP" has appeared on signposts and in Greenspeak (written in the windows of the Green Building). It has even been printed on shoelaces. The point is to use it creatively: I Hate To Face Physics, It's Hard To Fondle Penguins, I Have Truly Found Paradise. And of course there's the age-old tuition gripe, I Have To Forever Pay. But these flights of fancy are merely riffs on IHTFP's widely accepted primary meaning: I Hate This F*&^ing Place.

A familiar MIT acronym appears on the Great Dome, 1995.

GREENER PASTURES: THE GREEN BUILDING HACKS

MIT's Green Building is a hacker's dream. It is tall—one of the tallest on campus, rising 23 stories to 277 feet. It has ample display space—a symmetrical grid of more than 150 windows. It is highly visible—from both Boston and Cambridge. It even has a massive radar dome (or radome) that sits atop the flat roof. Not surprisingly, even before construction was finished in 1964, hackers had already staked their claim by suspending a "Tech is Hell" banner from the top of the construction site's pile driver.

Over the years, hackers have dressed the Green Building's 26.5-foot radome in various guises. In 1983, they fashioned it into a smiley face using giant yellow building tarps. In 2001, they turned the dome into a Magic Pi Ball in honor of spring finals. A spoof on the Magic Eight Ball (a perennially popular toy for predicting the future), the Magic Pi gave a quintessential Eight Ball response to students who sought exam advice: "Outlook hazy. Try again."

Because of its visibility from the Charles River Esplanade, hackers have often mounted Green Building hacks to entertain the enormous crowds that line the banks for the annual concert and fireworks on Independence Day. In 1986, when the Boston Pops was lured to New York City to celebrate the rededication of the Statue of Liberty, Bostonians were left without their traditional concert on the Esplanade. Hackers resolved that a trade was in order—the Pops for the statue. They transformed the radome into Lady Liberty herself standing watch from her high perch over the Esplanade.

The idea was simple, the execution not so. Hackers arrived with an extension ladder for reaching the hatch on the radome but quickly realized the ladder would not fit in the elevator or up the stairs. Using the 300 feet of rope they had brought along—just in case–they hauled the ladder up the side of the building. A hacker wearing a safety harness then mounted the radome and guided the network of spikes around the lightning rod. The crown was actually constructed of aluminum masts sheathed in white cloth and mounted on a wooden base, its underside lined with carpeting to protect the radome. Guy wires kept it in place. With the artful positioning of a yellow sheet, the crow's nest that rises beside the radome was pressed into service as Liberty's torch. Thus, a sense of civic balance was achieved as the Pops serenaded New York.

A few years later, on July 4, 1993, hackers turned the building into what they allege to be the world's largest sound (VU) meter. Hackers converted the ventilation ducts across the top of the Green Building into a VU meter. Created from 6-by-4-foot apertures that were illuminated by bright red lights, the 5,000-plus-watt meter was, at 250 times the size of an ordinary stereo sound

The radome is transformed into a smiley face using yellow tarps, 1983.

meter, the largest in the world. The light show was keyed to the music of the Pops concert (which didn't quite work because sound travels more slowly than radio waves), and later to the sounds of the fireworks (which did work well). The delighted crowd watched as "Cylon" scanning light patterns (reminiscent of *Battlestar Galactica*) alternated with one-dimensional Tetris and even "IHTFP" in Morse code.

GREENSPEAK SPOKEN HERE

To MIT hackers, one of the Green Building's most inspiring characteristics is its fenestration, which they have turned, time after time, into an enormous message board. In the Institute vernacular, this genre of hacks has come to be known as Greenspeak—the art of turning on and off lights and raising and lowering window shades to display a message.

The first recorded instance of Greenspeak occurred in March 1964, before the Green Building was even completed. The Theta Chi fraternity used the medium to increase their visibility, pulling off a giant "ΘX." Not to be outdone, the venerable hacking group Jack Florey used Greenspeak the next day to flash that iconic acronym of MIT life, IHTFP. Over the years, hackers have used Greenspeak to celebrate big moments in the life of the campus and the cities of Cambridge and Boston.

MOONWALK, 1969

When the Apollo 11 mission delivered Neil Armstrong and Buzz Aldrin (1963 ScD) to the moon's surface, hackers celebrated in Greenspeak, displaying a towering number 11 across the Green Building facade.

HOLIDAY LIGHTS, 1973 AND 1975

Hackers created a Christmas tree in Greenspeak, and two years later a jack-o'-lantern for Halloween.

EAST CAMPUS POINTER, 1979 THROUGH THE PRESENT

With a simple but attention-getting "ECÆ" the East Campus dormitory periodically uses Greenspeak to attract first-year students during residential recruitment.

MIT Salute to Apollo 11 Program

WELCOME — With the Cambridge skyline the background, a 10-story high numeral "11" in lighted windows of the MIT Faculty 's of the Charles River. Attending a party at the club were employes of the Technology laboratory. It was the lab where the Apollo 11 guidance and navigation system was developed.

Record American Photo. Frank Hill

Hackers salute the historic moon landing of Neil Armstrong and Buzz Aldrin.

Hackers created a Christmas tree on the Green Building, seen here from across the Charles River, 1973.

Hackers create a smiling jack-o'-lantern on the
Green Building for Halloween, 1975.

RED SOX VERSUS METS, 1986

Hackers added diplomacy to the usual kit bag of skills when Mets fans initially refused to pull down their window shades, nearly thwarting attempts to spell out "SOX" during the 1986 World Series. Ultimately the New York fans at MIT had their way, Greenspeaking "SUX" followed by "2" when the Mets prevailed over the Red Sox. There was much joy in 2004 when the beloved Sox won their first championship in 86 years.

GOOD WILL HUNTING, 1998

On Oscar® night, hackers spanned the fenestration to recreate the golden statuette in recognition of the Academy Awards® bestowed for *Good Will Hunting*, a film set at MIT.

Boston Red Sox fans' show of support for their team (left) is tharted by New York Mets fans (middle). Mets hackers wanted to make sure everyone knew that Boston had lost (right), 1986.

MAKING AN ENTRANCE: THE LOBBY 7 HACKS

Most visitors to MIT (and a gigantic portion of the MIT community) enter the Institute through the doors at 77 Massachusetts Avenue. This is the entrance under MIT's Little Dome. Technically, this is the Rogers Building, named for MIT's founder William Barton Rogers, but everyone calls it Building 7. The grand open space under the Little Dome is Lobby 7. Given the thousands who pass through Lobby 7 on a daily basis, it is not surprising that it is a popular venue for grand-scale *interior* hacks. Generations of students have passed through this gateway only to be greeted by welcome mats of bubble wrapping, an enormous rope swing, an intricate maze, or a giant chess board and on the prominent stone pedestals that frame the lobby, a student dressed as a military memorial or Winnie the Pooh.

In 1986, when graduate students in the architecture department's Building Technology Group suspended a space station prototype in Lobby 7, the Order of Random Knights hacking group wrapped it in 1,600 square feet of cloth, turning it into a massive six-sided die. In 2001, hackers created a faux archeological excavation site for the unearthing of a "large black monolith." They erected barriers and posted an authentic-looking sign showing a man in a chemical hazard suit digging up a massive black rectangle. This was just one of several allusions to the famous movie *2001: A Space Odyssey* that appeared on campus during the Class of 2001's four years on campus.

The hacks in Lobby 7 have ranged from the whimsically absurd—a downpour of 1,600 pink and green ping pong balls—to the monumentally surreal—a faux cathedral complete with stained glass windows, organ, and a wedding ceremony. A couple legally tied the knot during the 1992 hack appropriately titled Cathedral of Our Lady of the All-Night Tool. It is the only hack on record that enjoys that distinction, although a marriage proposal was delivered and accepted in a 2004 hack.

The Cathedral event, which took place on the night before Halloween and was carried out by a team of more sixty people, had been on the drawing board since the previous June, according to sources quoted in *The Tech*. Its creators had been inspired by a 1990 performance hack during which hackers from the Order of Random Knights dressed as monks and marched across campus chanting "Oh My God, Do We Need Sleep!" in Latin while handing out flyers for Our Lady of the All-Night Tool. In addition to intricate stained glass windows, mahogany pews, and massive stone tablets presenting the commandments, the hack included a computerized confessional manned by Father Eliza (a nod to former MIT Professor Joseph Weizenbaum), who ran on an Apple II+ and had to be rebooted when sins overwhelmed. (See "Father Tool's Grand Tour.")

Hackers wrapped a space station prototype hung in Lobby 7 to become a six-sided die, 1986.

Sixteen hundred pink and green ping pong balls were dropped over a period of two and a half minutes in Lobby 7, 1983.

The symbol from the popular movie *Jurassic Park* shines down from the ceiling of the Lobby 7, 1994.

Instead of a real oculus (the circular opening in classical domes that lets in light and air) at the top of the dome, anyone looking up in Lobby 7 sees a beautiful glass "skylight," even though the outside of the dome is solid. Hackers (and others) noticed that the "skylight" was artificially lit, indicating that there was a space between the ceiling of the lobby roof and the outer dome. If there is an interesting space on campus, hackers will find it, and this access point in Lobby 7 has facilitated some especially charming hacks, from the suspension of an enormous disco ball to an elaborate spider web, and even a lighted jack-o'-lantern. And the skylight's "occulus" has illuminated smiley faces, the Jurassic Park symbol, and Batman's official crest.

Two of the most ingenious hacks in the history of hackdom were executed on the inscription that rings the bottom edge of the dome at the fourth-floor level. In August 1994, hackers greeted incoming students with a bold, if subtle, welcome banner. They replaced the institutional slogan engraved in stone with one of their own. What once read:

Established for Advancement and Development of Science its Application to Industry the Arts Agriculture and Commerce. Charter MDCCCLXI.

now proclaimed:

Established for Advancement and Development of Science its Application to Industry the Arts Entertainment and Hacking. Charter MDCCCLXI.

The edit was accomplished with panels of polystyrene foam faux-finished to blend with the rest of the inscription and held in place by virtually imperceptible spring-loaded braces. The modification was so subtle, in fact, that campus police officers called to the scene to investigate suspicious activity found nothing out of the ordinary on their first two trips. Students and staff, learning of the hack, stared in puzzlement trying to figure out what had changed in the inscription. When finally they were able to distinguish the hacked panels from the original, the newly formed Confined Space Rescue Team from MIT Facilities rappelled down into Lobby 7 to remove it, determining that such a maneuver was the only way the hack could have been installed.

Three years later, celebrating Halloween 1997, hackers pulled off another stunning alteration of the Lobby 7 architecture, perching four enormous gargoyles on a ledge beneath the dome. Although fashioned of papier-mâché from hundreds of copies of a free newspaper distributed on campus, the 6-foot-tall statues gazed down onto Lobby 7 with eerie authority and authentic grotesquerie. Passersby, who previously had never had cause to look up, assumed they were stone statues original to the building. Closer inspection revealed, however, that one figure bore a striking resemblance to the cartoon character Dilbert and another to a beaver, the school mascot.

Cleverly painted polystyrene foam panels modify MIT's charter, 1994.

This close-up reveals the spring-loaded hardware used to suspend the foam panels above Lobby 7, 1994.

Four paper mâché gargoyles watching over Lobby 7, 1997. Bottom left: "Dilbert." Bottom right: "Beaver."

In addition to the hacks to the frieze and the addition of the gargoyles, there have been many delightful hacks under the dome. Among the most memorable:

THE STARSHIP *ENTERPRISE*, 1989

When William Shatner visited MIT in 1989, hackers prepared a surprise. A replica of Captain James T. Kirk's starship hovered above Lobby 7 as if pausing before accelerating to warp speed.

PAPER AIRPLANE, 1998

"Jolt is my copilot" was emblazoned on the side of this 30-pound "paper" airplane suspended above Lobby 7. Measuring almost 16 feet long and with a 7.5-foot wing span, this hack by the Guild of Dislocated Hackers coincided with the annual American Institute of Aeronautics and Astronautics paper airplane contest. An entry form for the craft is attached to a nearby pillar along with instructions for dismantling it.

SPIDER WEB, HALLOWEEN, 1980S TO PRESENT

Every so often a spider web will appear on campus—frequently, under the dome. Sometimes the web comes complete with spider bearing the name of an unpopular administrator.

JOLT CAN, 1995

Before our Red Bull era of hypercaffeinated drinks, the soft drink of choice was Jolt, described by the BBC website as the cola for "macho-nerds." With "all of the sugar and twice the caffeine" (according to its own ad campaign), Jolt Cola sustained many an MIT student through knotty P-sets (problems assigned for homework) and countless all-nighters. Just before final exams, hackers paid tribute to the drink of choice. The giant can "distributed by IHTFP Limited, Cambridge," bore authentic graphics and a list of ingredients: "Charles River Water, Sugar, Caffeine[2]."

PAPER DOLLS, 2002

After a long period of renovation, the "new" dome skylight was unveiled. Soon afterward, hackers accented the intricate glasswork with a ring of paper dolls.

WHEEL OF TUITION, 2002

On registration day for spring classes, the skylight was transformed into a wild game of chance with the spinner stopped at "Aid Denied."

MAKE WAY FOR CRUFTLINGS, 2003

This hack was inspired by the wonderful bronze sculpture of Mrs. Mallard and her eight ducklings featured in the beloved Boston-based children's book *Make Way for Ducklings*. A line of ancient computer monitors followed their digital "Momma Mallard" across Lobby 7. "Cruft" in MIT parlance means "obsolete junk that builds up in many a hacker's home or office."

Starship *Enterprise* in Lobby 7, 1989.

Large paper airplane in Lobby 7, 1998.

FATHER TOOL'S GRAND TOUR

Brothers and Sisters,

Let us take a guided tour of Cathedral Seven, the Cathedral of Our Lady of the All-Night Tool.

The first thing we notice is the striking stained-glass window that occupies the rear wall of our cathedral (west side, 77 Mass. Ave. entrance). It is composed of three distinct panels. The leftmost (as viewed from inside the Cathedral) depicts a common scene in the life of the devout tool— an MIT lecture hall, complete with students, blackboards, and professor. The center contains a rosette featuring the animal that walks most closely to the father above—a beaver (Brass Rat style). The rightmost panel shows the Killian Court view of the Institute, complete with Dome and a representative Tool—note that the Tool stands on top of a large stack of unmarked bills in order to gain entrance to the 'tute. The stained-glass window is quite striking when viewed from the side with the least light (inside during the day, outside at night). Also, when the father above sends us sunlight without clouds, the light shines through the glass and makes soft, clear images on the pillars (east side).

The stained glass windows in Cathedral 7, 1992.

Looking skyward, we see that the formerly dull skylight now radiates with light and color. In the central circle resides the logo of the goddess dear to many, an Athena Owl. Radiating outward from the center are pie-shaped wedges. There are 16 in all, colored alternately purple and blue. In each wedge is housed one of the many truths that form the foundation of our faith.

The skylight of Cathedral 7, 1992.

Of course, you all hear the inspirational music produced by our lovely organ. The present verse repeats hourly for all to enjoy. It starts with Bach's "Toccata and Fugue in D Minor" and continues with many inspiring pieces including some Gregorian chants. We can see the organ that produces this wonderful music on the second floor balcony (second floor, center east, directly at end of second-floor Infinite Corridor). Notice the ornate gold and silver pipes tuned for musical perfection. To take a closer look, we venture to the second floor. Here, we see the light from the middle window (Beaver) shines nicely onto the organ. The keyboard is constructed from six computer keyboards (four VT-100 and two which are believed to be of the Knight-TV system style). A close look at the keyboards reveals that important messages are contained within them as well (Nerd Pride, 37619*, IHTFP, Our Lady of the All Night Tool, 3.14159).

Organ in Cathedral 7, 1992.

Returning to the ground level, we note the row of pews on the south side of the Cathedral, facing east and the altar. The pews look quite like they belong in this Cathedral and it bemuses everyone. Few actually realize that the "pews" have always been in Lobby 7. A historical note—when originally installed, a different set of "pews" were used. The original set resembled lecture-hall seats. The pews were upgraded to their current status when a need elsewhere for the original pews became apparent. The altar has a "marble" top and has vintage 1971 4K core memory inlaid in its center. Also on the altar are a bottle of holy water (Jolt), a flask, and a holy tome (8.01 text). A gilded keyboard (Mac variety) sits at its base along with aluminum foil flowers (a donation from the wedding). The basic shape of the altar resembles an empty spool for wire.

Altar of offerings in Cathedral 7, 1992.

Behind the altar, we have the donation box where collection is being taken to cover the costs of Cathedral construction. The donation box is a VT-100 monitor which has had its insides removed. Its clear face allows one to see the donations being made.

On each of the long empty pedestals in the lobby, sit gilded artifacts and relics. On the SE pedestal are several gilded monitors (one with its innards exposed for all to appreciate). On the NE pedestal is a vintage TEK oscilloscope. On the SW pedestal is a PDP style 19" rack complete with a 9-track magnetic tape drive, punch-tape reader, and disk drives. The NW pedestal sports many artifacts including a Wang daisy wheel printer sporting dual disk drives, circuit boards, disk packets, and magnetic tape.

The location of the coffee pot at the Donut Stand is labeled "Holy Water." The picture of William Barton Rogers has a light shining upon it and is labeled "Saint Rogers." Over the entrance to the Infinite Corridor is a sign that proclaims this place of Holy Tooling, "Our Lady of the All-Night Tool."

On the NE and SE walls are two stone tablets displaying the 0x10 commandments:

Tablet 1 [Northeast wall]

0x0 - I am Athena, thy Goddess. Thou shalt not have false gods before me.

0x1 - Thou shalt not take the name of OLC in vain.

0x2 - Thou shalt not eat at Lobdell.

0x3 - Thou shalt keep holy the hour of *Star Trek*.

0x4 - Honor thy professors, for they are the source of grades.

0x5 - Thou shalt not decrease entropy.

0x6 - Thou shalt not connect PWR to GND.

0x7 - Thou shalt not sex toads.

Tablet 2 [Southeast wall]

0x8 - Thou shalt not exceed the speed of light.

0x9 - Keep holy the month of IAP for it is a time of rest.

0xA - HTFP.

0xB - Thou shalt not sleep.

0xC - Thou shalt consume caffeine.

0xD - Thou shalt not take pass/fail in vain.

0xE - Thou shalt not covet thy neighbor's HP.

0xF - Thou shalt not divide by zero.

Northeast Wall: First eight commandments of Cathedral 7, 1992.

Southeast Wall: The second set of commandments of Cathedral 7, 1992.

The donation box area of the Cathedral 7, 1992.

Adorning each of the central pillars at the east end of the Cathedral are two paintings. One depicts an adoration scene in which an HP28S is the object of adoration; the picture is complete with 1700s clothing, cherubs, scrollwork, and inscriptions "Hackito Ergo Sum," "Liberty, Fraternity, Equality, Caffeine," and "Novus Ordo Seclorum." The other depicts a man torn between Athena (complete with helmet, spear, and owl) standing on a platform inscribed "The Temptation of Six" and another woman standing on a platform inscribed "The Path of Truth"—in the background are two terminals with wings flying about.

Father Eliza provides the Cathedral with an eternally vigilant spiritual overseer. He is available for Holy Confession all hours of the day or night. Father Eliza has had special training in the needs of the MIT community. Father Eliza's confessional resides on the south side of the Cathedral and allows private conversations between tools needing guidance and the Father (Father Eliza is running on an Apple II+). Unfortunately, the magnitude of some people's sins is sometimes too much for Father Eliza and he must be rebooted regularly. We believe he is getting stronger as the days pass and he becomes better acclimated to the MIT community.

Mural of HP28S adoration in Cathedral 7, 1992.

Mural of Athena in Cathedral 7, 1992.

"ALL MONDAYS SHOULD BE SO BEAUTIFUL": THE ART OF HACKING ART

What many in the MIT community think of as one of the great Lobby 7 hacks was actually artwork and not a hack at all. The field of wheat that students and staff strolled through one Monday morning in May1996 was an art installation entitled "The Garden in the Machine." The nearly 100,000 stalks of wheat were planted by artist Scott Raphael Schiamberg (1993 SB, 1996 MArch, 1996 MCP), a graduate student striving to invoke the grand American pastoral tradition with an intimate, small-scale oasis.

Because of its sheer ingenuity, surreal impact, and obvious impermanence, the wheat field was seen as a colossal hack and proved irresistible to real hackers, who soon contributed a cow and a scarecrow to the pastoral scene. Schiamberg's "hack" was covered by the media, from the newswires to the networks, but the artist most valued the visitor feedback. Schiamberg's email inbox was flooded. "It took my breath away," one employee wrote, "All Mondays should be so beautiful."

The hackers' embellishment of the wheat field was a tribute rather than a critique, but hacks have more often served as a lively form of art criticism on campus. A common student lament is that MIT's taste in art is "big, black, and ugly." MIT's public art collection is critically acclaimed and includes many exceptional works (a few of which are, indeed, big and black). Two of the MIT community's most loved sculptures are Alexander Calder's, "Big Sail" and "Little Sail." Both have proven irresistible to hackers, and they have been adorned with smiley faces and other symbols. Once, in 1996, hackers added an authentic-looking plaque dedicating a fictitious companion piece, "The Great Wind."

For reasons that baffle the Institute's art historians and curators, Louise Nevelson's "Transparent Horizons" has become the campus sculpture that students love to hate. Students have claimed that the enormous black steel structure intrudes on the East Campus dormitory courtyard and that dorm residents were never consulted about the choice or placement of the art. For more than three decades, the two-story sculpture has been the butt of jokes, vandalized, *and* hosted a hack or two. Once hackers turned it into a study carrel with a desk and floor lamp set atop one of its highest planes. In a 1981 hack, the work was rededicated with a new plaque that read simply:

Louise Nevelson/b. 1900
Big Black Scrap Heap
1975

"Transparent Horizons" with a desk and lamp, 1984.

Calder Sculpture "Big Sail" with smiley face, 1996.

Art hacks, especially the addition of artifacts to the various campus museum exhibitions, have been a special penchant of one of the most enduring MIT hacking groups, James E. Tetazoo. Tetazoo's gallery enhancements have sometimes gone unnoticed (to the chagrin of curators) for embarrassingly long periods of time. For example, in 1979 Tetazoo installed a dime-store plastic aircraft carrier in a Hart Nautical Gallery exhibition along with a typical exhibit label with spoofed text.

USS *TETAZOO*

CONSTRUCTED IN 423 B.C. BY THE PHOENICIAN TURTLE KING SHII-DAWG, THE *TETAZOO*'S KEEL WAS LAID FOUR YEARS LATER IN DAMASCUS.

During the Middle Ages she was put into drydock in Norfolk, Virginia, until 1490 when she returned to Spain to show Christopher Columbo the route to the Americas under the new name "Ninny," later misspelled by Spanish hysterians. Running low on rum, she detoured to Puerto Rico where the wreck of the *Santa Maria* can be seen to this day.

In the early 1800s, she became a privateer under Sir Harry Flashman, C.A.P., C.I.A., C.O.D. Lost to the Swiss Navy in fierce combat in the Inside Straits, she remained in their possession until 1905 when she was given to the U.S. Navy as spoils from the Russo-Japanese war.

During WWII, she served with distinction in the Atlantic, sinking seven submarines, many of them enemy. Captained by James Tetazoo, Sr., she was named in his honor after he died while making a still from an old depth charge. To this day, she serves with pride as the only (official) floating still in the U.S. Navy.

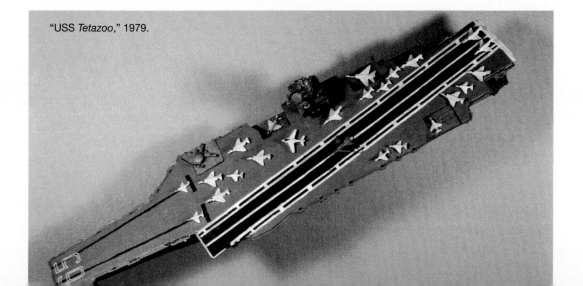

"USS *Tetazoo*," 1979.

In 1985, at MIT's List Visual Arts Center, Tetazoo added its own installation to a contemporary art exhibition. "No Knife," the overturned wastebasket with cafeteria tray and incomplete place setting was not recognized as a hack by gallery workers for several hours. "No Knife" was accompanied by a gallery label that attributed to the piece such artistically significant qualities as "temporary occasionalism," "casual formalism," and "sterile lateralism."

NO KNIFE
A STUDY IN MIXED MEDIA EARTH TONES, NUMBER THREE.
Realized by James Tetazoo, December 1984

The artist's mode d'emploi relies upon minimalist kinematic methods; space and time are frozen in a staid reality of restrained sexuality. Temporary occasionalism, soon overcome throughout by symbolic nihility, pervades our earliest perception of the work. An overturned throwaway obelisk functions as symbolic pedestal; the work rests upon a manifestation of gray toned absence. Epicurean imagery is employed most effectively by Tetazoo; the glass, the porcelain, the plastic move in conflicting directions and yet are joined in a mood of stark pacifism. The sterile lateralism of the grouped utensils (sans knife) conveys a sense of eternal ennui, framed within the subtle ambience of discrete putrefaction. The casual formalism of the place setting draws upon our common internal instincts of existential persistence to unify us with the greater consciousness of human bondage.

"No Knife," 1985.

But of all the art controversies in MIT history, the most heated has been the Great Hairball Controversy of 1990. MIT has a special program called "One Percent for the Arts." One percent of the budget for most new campus construction is set aside to underwrite a new piece of art. When the Stratton Student Center underwent renovations in the late 1980s, $75,000 was set aside to commission a new artwork for the lobby. MIT invited locally acclaimed sculptor Mags Harries to undertake the project. With the dual goals of involving students and evoking the essence of the Institute, Harries proposed a shaman's hat that would be fashioned from the hair of MIT students.

Harries divulged her underlying philosophy in the student newspaper *The Tech*: "Shamans were … the first scientists … the hats they wore, with four corners, which would be tied up together, were like a court jester's—the wise man and the fool speaking truths." Students promptly nicknamed the proposed sculpture: "The Hairball" and mounted a lively protest. The artwork was bound to be big and black and, many students felt, ugly too. According to the official plan, the sculpture would hang in the atrium between the first and third floors of the Student Center. Students lamented that the "hairball" would defeat the purpose of the renovations—to brighten the building and create an open atmosphere. Others were disturbed by the placement of the hair sculpture outside Lobdell, a student cafeteria. In addition to becoming the campus cause célèbre, the scheme inspired a spate of hacks.

When a steel frame prototype of the Harries structure labeled "Sculpture Testing" was suspended in the Student Center atrium, hackers placed a plate of birdseed under the cage-like structure along with the sign, "Acme Roadrunner Trap." Another hack cast the structure as a giant compass needle. When Harries' mock-up was removed, hackers suspended an eight-foot-long functioning slide rule in the same space along with the sign, "Alternate Sculpture Testing."

Just before an open forum to discuss the artwork, students organized a reading of "Green Eggs and Hair," a spoof of the popular Dr. Seuss book. Parodies of the Arts Committee statements and descriptions (aka "official hairball propaganda") were disseminated around the campus. Following the style of the administration's memos on the subject, hackers created their own "Big Questions" including "Why Ruin the Atrium?" In the end, the protesters triumphed—the hairball project never made it off the ground.

Mags Harries created a steel frame study piece for her proposed student center sculpture. Hackers cleverly draped it to become a bird, 1990.

When Harries' study piece was removed, hackers introduced a giant slide rule labeled: "Alternate Sculpture Testing," 1990.

Why ruin the atrium?

The Student Center Renovation Project was expensive--millions of dollars were spent transforming a dark, gloomy pit into a bright, trendy shopping mall.

Why spend $75,000 of the renovation fund to ruin the big wide-open "soul" of the building?

If Mags Harries really wants to find the proper place for this sculpture, maybe she should ask her friends over at the List Visual Arts Center if she can hang it in THEIR atrium. Students probably wouldn't mind the loss.

If Julius Stratton were alive today--and he is-- he probably wouldn't like to know that the quotation plaque of his in the atrium will be partially obscured by the sculpture, and high above his bust a large hairball will hang.

Of course, it wouldn't be fair not to mention the conspiracy theory. That is, it could be that ARA is secretly supporting the hair sculpture so that students will have something else to blame for hair in the food at Lobdell...

You will eventually stop complaining...

Green Eggs and Hair

I am art.
I'm made of hair.

That piece of art
That ball of hair
I do not like
That ball of hair.

Would you like me in Lobdell?

I would not like you
in Lobdell.
I do not want you
if you smell.
I do not want you
so you see,
I do not want you
over me.

Would you like me where you eat?

I would not like you
where I eat.
I do not want you
on my meat.
I do not want a ball of hair.
I do not want you anywhere.

Would you like me as a hat?

I do not want a
shaman's hat.
No superstition—
Nothing like that.
I do not want you so you see.
I do not want you at MIT.

Would you like me in your atrium?

In our atrium?
How could you dare?
You'll block the light
With your hair.

Fine. Fine.
Your point is clear.
You do not like me.
So I hear.
But I am art,
And I will be.
You will like me—
You will see!
(pause)
Would you like me
as a slide rule?
Thin and clear,
the engineer's tool?

Say!
I like a slide rule,
Thin and clear.
An object used
By the early engineer.
You're not superstitious,
You're not made of hairs.
And I do so like you
Over the stairs.
I do so like you,
And now I see,
That I would not mind you
At MIT!

FORM + FUNCTION = HACK: THE ARCHITECTURE HACKS

MIT students have found hacking to be as effective for poking fun at campus architecture as it is for art. In 1988, for example, when the plans for the Stratton Student Center renovation were posted, hackers overlaid the rendering with a print of an M. C. Escher staircase that bore an uncanny resemblance to the actual design. In the same year, with a few cans of red, yellow, and green paint, they turned a vertical row of round windows in the new Health Services building into a giant stoplight. (This was topped in 2007 by a hack that turned the windows into actual traffic lights.)

Stratton Student Center renovation plans with Escher influences, 1988.

The campus building that hackers seem to find most inspiring is alumnus I. M. Pei's (pronounced "pay"; S.B. 1940) 1985 futuristic box, the Wiesner Building, which houses the List Visual Arts Gallery and the Media Lab. As a harbinger of things to come, the dedication ceremony was met with a downpour of confetti that hackers had fed into ventilation ducts. Five pounds of shredded paper, 180 two-inch strips of magnetic computer tape, and 175 fish-shaped paper airplanes spewed out over the atrium—and over MIT President Paul Gray (1954 SB, 1955 SM, 1960 ScD), who happened to be standing directly under the ducts when the barrage was launched.

Before the building even was dedicated, the ever-whimsical James E. Tetazoo added a mint-green tile to artist Kenneth Noland's arrangement of oversized black, red, and yellow tiles adorning the building's facade. The extra square went unnoticed until the artist visited campus to view his work. A few years later, hackers "rearranged" the order of the squares by placing colored panels over them, confusing the Facilities crew when they arrived to remedy the matter. At one point, the facade bore two red squares and one black. And so it continues to the present day. This genre of hacks rarely attracts headlines, but MIT hackers love these subtle pranks that require specialist knowledge to detect and not so subtly suggest the intellectual prowess of the entire MIT community.

Not all hackers aim for sophistication. Some are just for fun, such as in 1986 when Jack Florey suspended a tire swing from the multistoried concrete arch that adjoins the building. On this same squared arch, hackers in 1989 installed a wide horizontal banner that read "Push Core for New Roll." (Inspired by its white tile exterior, students had dubbed Pei's architectural landmark "The Inside-Out Bathroom Building" and the "Pei Toilet," launching a multiyear spree of lavatorial hacks.)

In 1994, hackers created a bathroom stall, using the installation as an opportunity to parody other campus issues. The door and walls were covered with MIT-related graffiti and a magnetic card reader was installed for acceptance of the MIT Card, a debit card for students. Those using the Pei Toilet were instructed to run their cards through the slot to pay to enter and to exit the stall. An accompanying flyer described the toilet as a new Institute fundraising initiative.

On the tenth anniversary of the building's dedication, hackers took the lavatory theme a step farther with the 1995 Scrubbing Bubbles hack. Adopting the mascot of a popular bathroom tile cleaner, they installed the critters (enlarged to approximate scale) so that they appeared to be engaged in serious cleaning action on the white tiled exterior.

Perhaps it is I. M. Pei's spare aesthetic or that he is an alumnus—and it is likely just coincidence—but hackers seem to have a particular inclination to embellish his campus buildings.

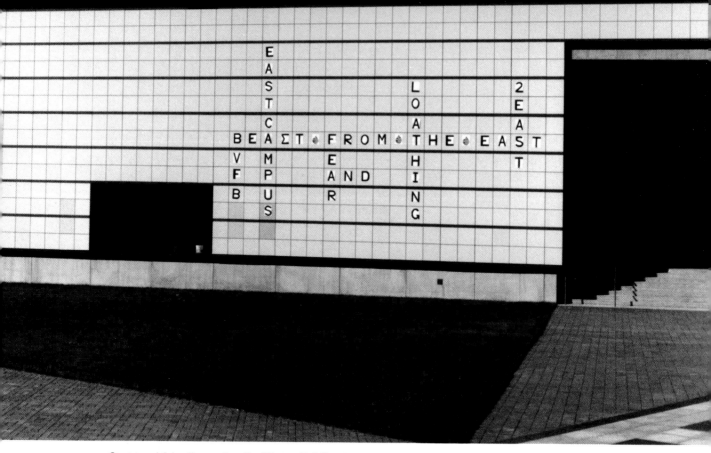

Crossword Advertisement on the Wiesner Building during freshman orientation week, 1986.

Students enjoy the tire swing hung from MIT's symbolic eastern gateway, 1986.

Three Scrubbing Bubbles on the exterior of the Wiesner Building, 1995.

The architect's design for Building 66, the Ralph Landau Chemical Engineering building, has inspired a succession of whimsical hacks since it was built in 1975. Simply put, the building's triangular shape and row of exhaust vent pipes on the roof creates the unmistakable silhouette of an ocean liner. During the building's dedication, residents of the neighboring East Campus dorm showed their appreciation for the resemblance by lowering an anchor over the building's "bow," unfurling a banner christening the building the "*USS Landau*," and blasting "Anchors Aweigh" on their sound systems.

Pei's "ocean liner" has been adorned with several anchors in its history. In 1992, spectators were particularly impressed with a gargantuan anchor dangling from a 30-foot chain off the building's starboard side. Closer inspection revealed the four-foot-long links were actually fashioned from black garbage bags over wire. In 1998, in a nod to the first campus showing of the movie *Titanic*, hackers christened the building the *Techtanic* and suspended a giant beaver over the helm.

One of the unique features of the MIT campus is its interconnectivity. The earliest plans for MIT's "new" Cambridge campus were shaped by John Ripley Freeman's observation that most of the time when students are in school, the weather in New England is poor, so why make them walk outside? When the campus began to extend beyond the original main group of buildings, one of the key design requirements was that the architects of the new buildings had to find a way to connect them to the old. Today it is possible to walk from the corner of Vassar Street and Massachusetts Avenue to the Sloan School of Management on the opposite side of campus without going outdoors. It is a convoluted network of bridges and tunnels that has served hackers well over the years. But once, in 1996, hackers took square aim at the colonial ambitions of Sloan School. Amid negotiations to take over parts of Building E40 (Muckley Building), hackers placed a sign with a giant X and the instructions, "Place Bridge Here" in the window of the third floor office suite that Sloan was hoping to take over.

More recently, the astonishing price tags associated with various campus buildings have inspired commentary by hackers. For example, on the morning of the dedication of the Ray and Maria Stata Center, designed by famous architect Frank Gehry, a gigantic MIT Property Office Sticker appeared on the white wall between the Gates and Dreyfoos towers. Parodying the twisted and skewed forms of the buildings myriad components, the sticker's barcode was twisted Stata-style (no right angles!). The serial number, MIT-285E06, reminded everyone that the estimated cost of the building was $285 million.

Anchor on the Ralph Landau Building, 1992 (homage to a similar hack in 1975).

Remembering the *Titanic*, the Ralph Landau Building is renamed the *Techtanic*, 1998.

A giant MIT Property Office sticker appeared on the new Ray and Maria Stata Center on the morning of the building's dedication, 2004.

CAMPUS COMMENTARY: HACKS AS POLICY PROTESTS (PLUS A FEW COMPLAINTS AND HUMOROUS SHOUT-OUTS)

MIT students view hacking as a handy forum on the whole gamut of administration policies and procedures. One perpetual beef is housing. Like students at many universities, MIT students over the years have considered the living accommodations to be too scarce, too small, too expensive, and generally below par.

In 1970, residents moving into the newly unveiled MacGregor dormitory were beset with plumbing and electrical problems in the unfinished building. Paying tribute to the building's contractor, hackers added a realistic cornerstone to the new facade. It was inscribed "Jackson Sux." In 1994, when the renovation of the Building 14 wheelchair entrance was repeatedly delayed because of administrative snafus, hackers adorned it with a package that read, "Do Not Open 'Til Christmas 2014."

Responding to an ever-tightening housing crunch that same year, hackers launched an elaborate prank. After freshmen registered their housing preferences, they received what appeared to be an official letter. There were several variations, but each meandered in diplomatic jargon to the eventual bad news, which was presented with keen bureaucratic spin. An excerpt:

Dear Freshperson:

You will be happy to know that this year we have given more careful thought to the needs and desires of the students most affected by these arrangements. We have hired several specialists in undergraduate housing psychology and, at great expense to the Institute, we have devised some very exciting new options for several students. You are one of the lucky few randomly chosen to have their requests withdrawn from the normal housing lottery to allow you to participate in this innovative new housing experiment. We are confident you will find this to be a delightful educational experience and a pleasurable way to start your student career at MIT.

Specifically, we have arranged for you to room with five other persons in a four-person police cruiser on top of the Great Dome. This living arrangement will provide you with a magnificent view of downtown Boston. Your accommodations will include free coffee and doughnuts, provided daily by Campus Police via their traffic monitoring helicopters. Conveniently, this living arrangement provides quick access to 10–250, one of the most frequently used lecture halls for freshperson classes. You will also be glad to hear that, in winter months, you will be provided with gasoline to fuel the car for heating purposes. As an additional benefit, after the first month, you will be waived the $500 roof access fine.

Sincerely,

Jack Florey, B.S.
Dean of Freshperson Housing Assignments

Institute housing administrators worried that some first-year students might either take these instructions literally *or* decide to go along with the game and head to the Dome to deposit their worldly belongings. They rushed to follow up with the official assignment letter, which began:

No, you haven't really been assigned to live in a police cruiser, a freight elevator, or the Chapel! If you received a message claiming this, or something similar, you've just encountered an MIT tradition known as a hack.

Two years later, hackers celebrated Halloween with another commentary on the housing situation. In an empty museum display case in the in the student center, they created a modular dorm room, complete with furniture, books, soft drink, Chinese food, and student dummy sporting a propeller beanie. A proposal posted on the outside of the 48-cubic-foot cube touted the "reengineering-inspired" Housing 2000 project as "a new housing paradigm to take MIT into the next century."

The escalating price of tuition has been a contentious issue, virtually since the day the Institute opened its doors, so it stands to reason that the spontaneous tuition riot is one of MIT's most enduring traditions. The "riot," which often takes place on the day the rate hike is announced, consists of a motley collection of students wielding colorful signs along the lines of "$25,000 and Still One-Ply Toilet Paper!" or the simple "$25,000 is TDM" (too damn much), a perennial favorite.

On the hallowed day in 1993 when the Institute announced that a single semester's tuition was hitting the $10,000 mark, hackers "adjusted for inflation" the 13-by-30-foot one-dollar-bill mural on the MIT Cashier's Office. The hallmark of a good hack is painstaking detail and the

Housing 2000 Project in the student center, 1996.

$10,000 bill employed authentic markings from the 1918 series, including the likeness of Salmon P. Chase, father of the National Banking System (the government stopped printing $10,000 bills in 1946). Thanks to the application of a few basic engineering tricks, the original mural was not damaged in any way. While it appeared to replace the one-dollar bill, the new denomination was reproduced on stretched muslin mounted imperceptibly over the existing mural.

Hackers in 1998 used a less subtle approach to make their point about tuition inflation, greeting the incoming classes with a voluminous banner, "Welcome MIT Debtors."

$10,000 bill outside of the Cashier's Office, 1993.

Cafeteria fare is a common student complaint on most college campuses, and MIT is no exception. All of the campus food-service providers have been hacked from time to time, but 1995 was a particularly fruitful year. Inspiration came from two sources—the opening of the Bio Café in the new biology building and Aramark's penchant for gimmicky dining themes. On spring registration day, when MIT services traditionally set up information booths to answer questions and distribute materials, hackers set up a booth allegedly representing Aramark's most recent dining innovation—the Biohazard Café. Hackers in lab coats and chemical clean-up suits staffed the booth, passing out menus and encouraging visitors to sample dishes named Soylent Green, Lab Mice on Rice, and Humungous Fungus Pizza—provided they signed the release forms.

From Drosophila Delight to MITochondria, the hacker's Biohazard Café added something special to MIT's eating options, 1995.

MIT president Jerome Wiesner (1971–1980) famously said in a speech to students that "Getting an education from MIT is like taking a drink from a fire hose." The expression caught like wildfire and students have reproduced it on banners and T-shirts ever since. During final exams in 1991, hackers converted a fire hydrant into a fully functioning drinking fountain outside Room 26–100 (the largest lecture hall on campus) so that students could literally "take a drink from the fire hose." Hackers re-plumbed the fountain so that pressing the fountain's lever caused water to flow through the hydrant, and come out the fire hose nozzle suspended over the fountain and replacing its spigot.

Fire hydrant drinking hose, 1991.

GETTING AN EDUCATION FROM MIT IS LIKE TRYING TO GET A DRINK FROM A FIRE HOSE.

But the MIT administration hasn't been the only target of such protest hacks. Occasionally, hackers have been inspired to prepare a special welcome for auspicious Institute guests. When

Al Gore Buzzword Bingo!

Welcome to Buzzword Bingo! For the past four years, most of you have enjoyed as a spectator the fine tradition of hacking at MIT. Today, as you finish your time here at MIT, you will have a chance to enjoy it as a participant. Soon, Vice President Gore will deliver an address on Distributed Intelligence. We will greet him with a Distributed Hack. Like any good distributed system, there will be no single point of failure, no single person whom the Campus Police or the Secret Service can stop.

As MIT students, and now as graduates, you are surely familiar with the tendency of non-technical people to use buzzwords when discussing technical issues. The Vice President, although more technically aware than most of his colleagues, is sure to use this technique in his speech. This hack is designed to gently remind him that he is at MIT, where we can see right through this strategy.

Below, you will find a Bingo board. This is similar to a regular bingo card, except that each square contains a buzzword instead of a number. When Al Gore uses a buzzword on your board, cross it off. If you get five buzzwords in a row - horizontally, vertically, or diagonally - you have won Buzzword Bingo! Instead of shouting "Bingo!" (which would be rude and potentially upset the men with wires in their ears and guns all over the place), hold up your card so that the other side faces the podium and the Vice President can see that you have won.

Have fun!

information superhighway	National Challenge	human-responsive	bitway	megaflop
Global Information Infrastructure	distributed	assimilation	vector	infobahn
empower	Information Age	FREE SQUARE	milestone	knowledge worker
NSF	communications	methodology	framework	environment
interoperability	virtual	environment	information space	mission goal

board id number = 1149160

Al Gore Buzzword Bingo!, 1996.

former vice president Al Gore gave the commencement address at graduation in 1996, he soon noticed that the assembled masses were diligently marking Bingo cards as he spoke. In fact, they were playing a game that hackers had distributed: Al Gore Buzzword Bingo! Instead of numbers, each square contained a familiar Gore buzzword. Whenever Gore uttered a word like "infobahn," "hybrid," or "paradigm," the students would check it off. "If you get five buzzwords in a row—horizontally, vertically, or diagonally—you have won Buzzword Bingo!" the instructions read. "Instead of shouting 'Bingo!' (which would be rude and potentially upset the men with wires in their ears …), hold up the card so that … the Vice President can see that you have won." Gore had been tipped off to the game, apparently, because when students cheered at one point in the speech, he quipped, "Did I say a buzzword?"

Microsoft has also seen its share of MIT hacks. In 1994, a hacking commentary on the software giant was as ubiquitous as Microsoft itself. Signs were posted throughout the main Institute thoroughfare using official Microsoft fonts, graphics, and logos with messages that made it seem the software empire controlled the world, from windows to doors, vending machines to toilet seats. Signs on stairways, for example, read "Microsoft UpwardConnect™ for Walkgroups vO.9Beta (not yet downward compatible)," and vending machines were labeled "Microsoft Dispens-O-Fud™ for YOU v32pi/3 (Think of our food as stock in our company, and invest your money here!)." When Gates visited two years later, hackers stretched enormous banners across some of the most prominent facades on campus. "MIT doesn't do Windows," read one. Another replicated the Windows95 signature "Start" button, substituting the word "Crash."

Windows protest banner, 1996.

A SIGN OF THE TIMES:
HACKING WITH SIGNS AND BANNERS

The MIT community awaits hacks on April Fool's Day with the same anticipation that weather watchers look for Punxsutawney Phil on Groundhog Day. But when the campus population looked up expectantly at the Great Dome on April 1, 2000, they saw instead an enormous banner stretched across its august pedestal that read "Ceci n'est pas un hack" ("This is not a hack")—a wry play on Magritte's *The Treachery of Images*.

Signs and banners are indeed one of the mainstays of hacking. In 1925, *The Tech* included a short note that "The boys in '93 Dormitory are certainly getting high hat as to signs." An enormous electric sign blazed: "Suffolk County Jail." "What next," asked *The Tech* writer, "the Station 16 sign? The Lounger would not be surprised to find the gilt from the State House Dome transferred some night to the big dome of Building 10."

Ever since, hackers have posted or transformed hundreds of signs. On one night in November 1987, for example, the large sign outside the Student Center advertising Starvin Marvin's Café at the Sala de Puerto Rico became "Salmonella de Puerto Rico Café," complete with a graphic of a squirming bacterium.

MIT's tradition of numbering its buildings has likewise invited many spoofs. In 1991 hackers decided to erect their own sign for the upcoming dedication of the new Building E40. The authentic-looking plaque renamed the new addition to campus the "E. Phortey Building" in memory of "Edwin Phortey" and thanked "W. D. Phortey '40 and N. Dallas Phortey" for the funds necessary to make the building a reality.

In 1996, hackers replaced the regulation "No Trespassing" signs at the entrance to computer clusters with an edited version that appeared to be identical to the original—identical, that is, except for the addition of a graph and the message, "You must be at least this smart to use Athena workstations." The graph charted levels of intelligence in increasing increments starting at the low point, "Urchins who log in as root" and working up past "average Harvard student," "average B.U. student," and "average CalTech student," to the high point on the scale, "below-average M.I.T. student."

During renovations in the AeroAstro department in 2000, some employees relocated to temporary facilities across from the building on Vassar Street. Hackers decided to boost the morale of those interned in the dismal mobile complex by posting a new sign on the chain link fence outside: "Aero-Astro Estates, A Trailer Community. It has been 71 days since our last tornado." The tornado count was updated regularly during the months the sign remained in place.

Details and authenticity are as essential in signs as they are in any hack. For that reason, the 1987 Nerd Xing sign is, in many ways, the gold standard of sign hacks. The bright yellow crossing sign, posted at the busy crosswalk at 77 Massachusetts Avenue, cleverly morphed a textbook nerd into the regulation graphics of a municipal crossing sign. Connoisseurs appreciated the details: the regulation nerd backpack, a "nerd kit" (the briefcase like container holding a lab kit that serves as an experimental platform for prototyping digital and analog projects), and the storage medium of choice for 1987—a 5.25-inch floppy disk.

Hackers transform a crossing sign into a "Nerd Xing" sign, 1987.

Then there are the banners welcoming freshman. Hackers routinely charm the incoming students at the annual picnic. "Abandon All Hope Ye Who Enter Here," hackers warned in 1975, quoting the inscription on the gates of hell in Dante's *Inferno*. At the 1979 picnic, the front of Building 7 quoted Poe: "For the Love of God, Montressor," the last words spoken by the character in "The Cask of Amontillado" as the final brick is put into place, sealing him into a wall forever. Right below this banner, a second group of hackers hung a sign reading: "Cambridge Tool Co." Apparently, the two groups were unaware of each other's plans.

Two hacks are unfurled at the same time. On top, hackers quote from the closing line of "The Cask of Amontillado." Below, is a banner renaming MIT the "Cambridge Tool Co.," 1979.

In the entire panoply of sign hacks, one that still impresses as the decades pass is the Sheraton Boston Hotel Sign. Back in the fall of 1967, members of the Alpha Tau Omega (ATO) fraternity decided they needed to up the ante on marketing during residence/orientation week (R/O). That was the period just prior to the start of classes in the fall when dorms, frat houses, and other living (housing) groups vied for first-years. (Since 2001 all first-year students have been required to live in dormitories.) A cadre of ATO frat brothers found their way to the mechanisms controlling the Sheraton Boston sign, a prominent landmark on the urban skyline positioned directly across the Charles River from MIT. With the flip of a few switches and the aid of large black oilcloth, the hackers darkened all the letters except three—ATO. Several times since then, the Sheraton Boston obligingly blackened its letters for ATO one night during every R/O.

Alpha Tau Omega uses the Sheraton sign to advertise during Freshman Orientation Week, starting in 1967.

But the power of most sign hacks is that they speak for themselves. The images in this section illustrate the broad range of banners and signs—silly and serious—that hackers have deployed over the years.

M.I.T. monster Eats Boston/Back Bay sign hack, 1985.

Hackers question MIT
students' ability to walk
and chew at the same time
in the Chew Crosswalk
Signal sign hack, 1977.

Department of Alchemy,
pre-1960.

Gnurd Crossing sign, early 1970s.

Tropic of Calculus door sign, 1987.

Freshman Picnic sign,
R/O Center Baggage Sale,
1983.

Heaven and MIT via
Elevator, 1980s.

THE NUMBERS GAME:
HACKERS REINVENT MEASUREMENT

Numbers loom large in the MIT culture, so it's not surprising that hackers feel the need to work them from time to time. On April Fool's Day in 1972, for example, they published a handy manual called the "Alphabetic Number Tables" that spelled out and alphabetized all numbers 0–1000. "Availing ourselves of the unmatched technological facilities of this Institute," the introduction read, "we have developed, compiled, and revised these listings in the hope of bridging the cultural gap separating theoretical investigation and practical application." They hawked the manual for fifty cents in Lobby 10.

The same year also saw the birth of the Bruno measurement. The Bruno is the unit of volume equal to the size of the dent in the asphalt resulting from the six-story free fall of an upright piano. On October 24, 1972, the Bruno was 1158 cubic centimeters. Documented with microphones and high-speed movie cameras, the experiment revealed that the piano was traveling at 43 miles per hour and had 45,000 foot-pounds of energy at impact. The unit is named for the student who suggested the experiment as a means for disposing of an old dorm piano, a campus tradition that has been repeated over the years whenever an appropriate (i.e., beyond repair) piano is available. (Some consider the original Baker House Piano Drop to be a hack; some do not. All agree that today the event is considered a tradition.)

In 1997, hackers decided that the Institute's Infinite Corridor established a mathematical precedent. After all, the hallway—thought to be the longest straight corridor in the world—has been called "Infinite" for as long as anyone can remember. And since it is 47.2 rods long, it stands to mathematical reason that infinity must be 47.2 rods long. Thus it was expressed, deliberated, and converted to other units of measurement on posters and banners all along the Infinite Corridor.

Measuring a Bruno, 1972.

RECALCULATING THE INFINITE CORRIDOR

According to the signs posted along the Infinite Corridor during the 1997 Implications of Infinity hack, if the Infinite Corridor really is infinite, then …

The Harvard Bridge is approximately 120 rods across, so one must travel 2.5 times an infinite distance to walk across it. This might explain the attendance records of people from some of the Independent Living Groups across the river.

SMOOT POINTS

The greatest measurement hack of all time is indisputably the Smoot—named for Oliver Smoot, the Lambda Chi Alpha fraternity pledge used to recalculate the length of the Harvard Bridge in 1958. It seems that Lambda Chi fraternity brother Tom O'Connor got it into his head to give his new pledges a chance to do public service. He figured that in making the long trek across the bridge from Boston, where the fraternity was located, some indicator of progress would be helpful when frat brothers held their heads down against the rain, sleet, snow, and gale force winds they sometimes encountered en route.

Smoot, at just under 5'7" was the shortest pledge among the freshman at hand and was thus elected to be the human yardstick. Using swimming-pool paint on the bridge sidewalk, the pledges marked every Smoot with a colorful tick mark and at every 10 Smoots spelled out the full measurement. Eventually, they reached the end of the bridge—at 364.4 Smoots, plus or minus one ear.

Every two years, Lambda Chi pledges repaint the markings, which Cambridge police officers have come to rely on for specifying exact locations when filling out accident reports. After graduating from MIT in 1962, Smoot attended Georgetown University Law School. The man who is perhaps the most famous human-based measure of the twentieth century served as chairman of the American National Standards Institute (ANSI) and president of the International Organization for Standardization (ISO). By some serendipitous twist, the progeny of both Smoot and O'Connor enrolled at MIT's Sloan School of Management forty years later.

The Alumni Association marked the fiftieth anniversary of the Smooting of the bridge in grand style: the City of Cambridge declared October 4 to be Smoot Day and the Institute dedicated a special plaque to commemorate the occasion. The MIT students who designed the plaque also created a Smoot stick, a permanent tool that will serve as the standard Smoot measure far into the future for MIT students who want ensure the accuracy of the length of the Harvard Bridge.

Brothers of Lambda Chi mark out Smoots on the Harvard Bridge, 1958.

At start of the Harvard Bridge in Boston, the Smoot countdown (364 +/- 1 ear) begins. . . .

The halfway marker across the Harvard Bridge, as measured in Smoots.

Oliver Smoot holds the plaque commemorating the fiftieth anniversary of the Smooting of the Harvard Bridge, 2008.

BEYOND RECOGNITION: COMMEMORATION HACKS

Although the MIT community does not suffer fools gladly, it does have a sentimental streak. Students rarely let important (and even some less-than-important) occasions pass without observing them in some fashion—greeting William Shatner with scale model of the Starship *Enterprise*, for example. Some of the most memorable hacks over the years have commemorated holidays, celebrations, official visits, or just the start or end of school.

Hackers have contrived all sorts of ways to initiate newcomers into MIT culture. As MIT President Paul Gray gave his annual speech welcoming incoming students in 1986, Kresge Auditorium was suddenly filled with the strains of Bach's "Toccata and Fugue in D Minor"—this might have been a pleasant diversion except for the unnerving fact that no one was seated at the pipe organ in the mezzanine.

In August 1990, a photographer was organizing the incoming students on the colonnade just below the Great Dome for the Class of 1994 photo. After arranging and rearranging the crowd of students, he took aim. Suddenly, a vertical banner unfurled from the entablature high above. "Smile," it read, immediately followed by a deluge of 1,994 smiley-faced superballs. The balls bounced 20 or 30 feet into the air after the first impact just behind the assembled students; soon the air was alive with ricocheting spheres. None too pleased, the photographer shouted for attention, "Please stop throwing balls at the camera!" and then ducked the resulting onslaught.

At the annual frosh picnic in 1994, hackers opted for a variation on that theme to initiate the Class of 1998. As the official Institute spokesman welcomed the incoming class, a 12-foot balloon slowly inflated on the roof of Building 10 just below the Great Dome. Constructed of hexagonal and pentagonal plastic sheets taped together, and inflated with a leaf blower, the burgeoning balloon was later described as having a "fullerene configuration." To the assembled students it looked like the world's largest soccer ball. The hexagons were painted with the letters "MIT," the pentagons with "IHTFP." (The hack was probably also a nod to the World Cup Soccer championships that had been hosted locally over the summer.)

More than a decade before, in 1986, organizers of the frosh picnic schemed to sidestep the hacking distraction by moving the speeches to the more controlled environment of Kresge auditorium. Only first-years would be allowed. Alas, a strategically positioned hacker infiltrated the auditorium and directed more than fifty first-years to blow bubbles on cue during the president's speech. Soon the auditorium was filled with bubbles. As the luminaries seated on stage batted away the bubbles that were drifting their way, President Gray finished his prepared remarks.

Smile banner accompanied by 1,994 smiley-faced superballs, 1990.

MIT/IHTFP ball fully inflated, falling off of the roof, 1994.

MIT presidents have long been the target of hackers. In 1990, when President Charles Vest reported for duty on his first day, he had the unsettling sensation of having misplaced his office. Vice President Constantine Simonides, who was escorting Vest, was even more dismayed, having a long knowledge of that particular hallway. Where they both believed the door should be, however, hung an enormous bulletin board covered with clippings hailing Vest's arrival. They soon realized that hackers had considered it their duty to commemorate the occasion.

Vest had come to the Institute from the University of Michigan in 1990, but his actual inauguration took place many months later, in May 1991. A hack would have been conspicuous by its absence at such an auspicious event, and hackers knew it. Inspired by Michigan's myriad satellite schools, they strung an enormous banner across the Student Center facade that read, in the official UM font and colors, "University of Michigan at Cambridge."

Whether turning columns into candy canes at Christmas or installing a shower stall in the computer cluster on April Fool's Day, hackers can be counted on to get into the spirit of holidays and celebrations—or invent them when necessary. During final exams in December 1994, hackers decided the time was never better to celebrate "Vector Day." The "Welcome to Athena" login banner on computer workstations at the Institute was replaced with the greeting "Happy Vector Day." Everywhere on campus, banners celebrated the holiday, and large vectors mounted on walls and over doorways pointed this way and that. In the vector calculus class, hackers distributed a commemorative assignment.

Commencement is a natural opportunity for hacks. Along with the 1996 Al Gore Buzzword Bingo! prank, one of the most memorable graduation hacks in recent years was the 2001 Parachuting Beaver hack. Graduates sitting in Killian Court on that fine June day suddenly spotted a huge balloon drifting overhead. Shortly afterward, the sky was thick with parachuting beavers (the plush toy variety, of course). Carried off by the balloon, the deployment mechanism washed up on Cape Cod a few days later.

Some of the most effective commemoration hacks have marked the passing of something or someone beloved. When MIT finally made preparations to tear down the barracks-like Building 20, which was built as a "temporary" facility during WWII, students recognized the demise of the fifty-five year-old structure with an official-looking but surreally oversized "Deactivated Property" sign. The birthplace of many groundbreaking innovations during its years, Building 20 was alleged to house a secret subbasement where great minds convened to change the world. After the building was torn down, hackers paid tribute to that legendary subterranean think tank by erecting an elevator on the leveled site, lending the impression that the hidden laboratories were still down there somewhere. In fact, there is a persistent joke on campus that the building is still standing but hidden by an invisibility shield.

When the Institute put an end to the thirty-three-year-old Rush tradition, hackers observed its passing with a variety of hacks. Rush was the period during which incoming students chose where to live. During this time, they were courted royally by dorms, independent living groups (ILG), and fraternities and sororities. In conjunction with sweeping changes relative to student housing, however, the practice was modified in 2001 when all first-year students were required to live on campus. For the ILGs and Hellenic groups that had once provided the majority of housing for incoming students, this was a very dramatic change.

That fall, hackers solemnly eulogized the old Rush. Under the Great Dome, they hung a wide, wistful banner that, in MIT fashion, also referenced the late geek-culture hero Douglas Adams: "So Long and Thanks for All the Frosh." The banner was positioned to be a prominent feature in the annual class photo, which was happening on the steps below. New students must have wondered what they had missed with the passing of Rush, because as they mugged for the photographer, a funeral procession then emerged across Killian Court. A trumpeter playing "Taps" headed the procession followed by a group of students dressed in black. They in turn were followed by pallbearers carrying a massive tombstone inscribed "RIP RUSH 1968–2001." The mourners held a brief service then moved on, leaving the tombstone behind for the new students to puzzle over.

The "Fishbowl" was a communal computer cluster of twenty-nine workstations, named for its large plate glass windows that overlooked the Infinite Corridor. When the MIT administration decided in 1997 to relocate the cluster as part of a consolidation of student services, hackers organized a formal lament. First, with a few pieces of jagged plexiglass , they gave the impression that a rock had been thrown through the Fishbowl and that water had spilled into the hallway. Paper fish appeared to flounder among the puddles. Wet footprints tracked to the Student Services Office (the department held responsible for the move) and a plate of paper sushi was left at the door. The coup de grâce was the login symbol on Athena, the campus computer network— hackers replaced the standard owl icon with an ailing fish, its tail twitching in the final spasms of death. It being the impact-aware 1990s, hackers followed up with both computer and physical-plant staff on how to erase the hack, but officials took them up on their offer to put it right themselves. In a letter to hackers, the computer cluster supervisor noted that "Student Services has been reported to APCF, the Association for the Prevention of Cruelty to Fish."

Banner commemorating the end of Rush, 2001.

SO LONG, AND THANKS FOR ALL THE FROSH

Rush Funeral Procession, 2001.

RIP RUSH 1968-2001

151

Hackers were compelled to commemorate two somber events in 2001. A few days after the death of science-fiction writer Douglas Adams, they strung a banner under the Great Dome thanking him for his wit. A few days after the September 11 terrorist attacks, they posted a massive 25-by-16-foot American flag in the same spot.

THE CASE OF THE DISAPPEARING PRESIDENT'S OFFICE
Charles M. Vest, MIT President, 1990–2004

It simply isn't even close when it comes to naming my favorite and most unforgettable hack. That's because I was the hackee.

It all goes back to my very first day on the job … on Monday, October 15, 1990. The late Vice President Constantine Simonides was escorting me to my office—his was across the hall—on the second floor of Building 3. When we arrived, however, there was no office to be seen, only a large bulletin board, flush against the wall and covered with newspaper clippings, including several about the search that lead to my selection as president, and also clips from *The Tech* headlines, "Vest Takes Over on Monday."

So good was the ruse that Constantine became momentarily disoriented and thought we had, perhaps, while engrossed in conversation, climbed the stairs to the wrong floor.

Then we broke out in hearty laughter when we realized what had happened. The bulletin board, ingeniously constructed and snugly fitted within the opening, was moved aside and, lo, there were the outer doors to the president's and provost's offices.

We gave the bulletin board a place of honor and humor for a time, and I still have it. As I explained later that day to a group I was addressing, "My first major policy is that we're going to keep that. The first time issues get hot on campus, we'll put it back in place." Well, there have been some fairly hot issues, but none so bad that I've had to hide behind the bulletin board.

If there was a message in all this, I suppose, it was that MIT presidents come and go, as do students, but the rich culture and traditions of the Institute will endure. The student hackers, who remained anonymous, left behind a bottle of champagne as a gesture of welcome and goodwill. Later, when we opened it, we toasted the hackers and MIT students generally, whose ever-inventive minds help to make MIT such a special place.

President Charles Vest finds his office entrance blocked by a bulletin board, 1990.

OBJECT LESSONS: HACKS IN THE CLASSROOM

Since the time in the 1870s that students sprinkled mild contact-explosives on the drill room floor, the MIT classroom has been a hotbed of pranks. One bright fall day in 1985, students filing into their physics lecture (8.01) took what they thought were the usual handouts at the front of the room. When they sat down and reviewed the day's assignment, however, they found that a class hack was afoot. The handout, parodying typical class assignments, supplied detailed instructions for the construction and launch of a paper airplane. The hack went off without a hitch at precisely 11:15, when hundreds of paper airplanes shot toward the lecturer.

Hackers deployed a mock assignment again in May 1992. Students showing up for their Structure and Interpretation of Computer Programs class (aka 6.001) realized one of the course handouts was a hack. Entitled the "6.001 Spellbook," it proceeded to detail a series of computer hexes that may have seemed all too real to many of the students in this challenging course.

During the 1992 presidential election, first-year students in the multivariable calculus class taught by Hartley Rogers Jr. decided that he was their man and launched a campus-wide campaign to get him elected. Playing on Rogers's pet phrase, they adopted the campaign slogan, "Hartley Rogers: The Intuitively Obvious Choice for President." During the campaign, students covered the projection screen above the lectern with campaign posters that remained hidden until Rogers lowered the motorized screen at the end of class. On election day, hackers deployed a remote-controlled car fashioned to look like the trolley from *Mr. Rogers' Neighborhood* that sped around the lecture hall accompanied by the strains of the Hartley Rogers campaign song, the "Mr. Rogers Show" theme. Though students knew their candidate would not receive a statistically significant number of votes, the musical tribute drew a standing ovation from the class. (Rogers did have a link to one of the real presidential candidates. His only foray into politics was when his fifth-grade classmate Barbara Pierce nominated him for class secretary. After elementary school, the two lost touch but Rogers recognized his childhood friend when her husband George H. W. Bush became active in politics.)

MASSACHUSETTS INSTITUTE OF TECHNOLOGY
Physics Department

Physics 8.01   October 25, 1985

ASSIGNMENT No. 7.1
(Do 11:15 a.m., Friday, October 25)

Reading Assignment for Week no. 7:
Review of airplane tossing technique, Chapter 22.5.
Advanced folding technique, Chapter 22.7.
Infra-red tracking and guidance techniques, Chapter 23.2.

Material Covered in Lecture this week:

Friday, October 25	Ready
Friday, October 25	Aim
Friday, October 25	Fire
Monday, October 28	First Aid

(special topic: treatment of paper cuts.)

Special Reminder:
The final exam in paper airplane construction and launch will be held (during the lecture time) on Friday, October 25. You will be held responsible for the material covered in chapters 22-3 (i.e. this homework.

Problem 7.1-1: Paper airplane construction.
If you don't know how to make one by now, you're probably beyond hope, but as Airplane construction is only a minor subtopic, I have included a diagram of the necessary folds as well as the final product. You may use this paper. Extra credit will be given for creative designs.

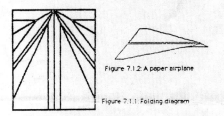

Figure 7.1.2: A paper airplane

Figure 7.1.1: Folding diagram

Problem 7.1-2: Paper Airplane Launching.
As its name suggests, this step involves the launching of the planes designed in problem 7.1-2. The goal is for the entire paper airplane fleet to impact the general vicinity of Professor Meyer at a given time. This involves proper launching technique to maximize the distance function with respect to force, and thus, the range of the airplane. Completing this problem will necessitate drawing a free-body diagram of the plane, including air resistance and lift, as well as a rapid measurement (to within 3 microns (1 micron = 1×10^{-6} m) accuracy) of the distance from your seat to the stage. As our goal is to minimize error, we must make many simultaneous trials of this experiment. Hint: throw it at the professor.

IMPORTANT NOTE: The launch is to take place at precisely 11:15 a.m. according to the room clock. Please have your equipment ready to by then. Fnord.

Physics 8.01 Paper Airplane instructions, 1985.

WHY SLEEP THROUGH A CLASS WHEN YOU CAN HACK IT?

SNAPPY ASSEMBLY, CIRCA 1870

Pranksters sprinkled iodide of nitrogen, a mild contact-explosive, on the drill room floor, lending snap, crackle, and pop to routine assembly.

PROGRESSION OF TARDIES, 1927

Although vowing that he would never shut the door in a student's face, a professor religiously locked the classroom at five minutes past the hour to discourage late arrivals. One day, students waited until five minutes after the bell, then trickled into the room at carefully spaced intervals so the professor could never close the door—at least not for another twenty-five minutes.

CASUAL SATURDAY, 1949

Students arrived at their early Saturday-morning class in robes and pajamas to protest the cruel and unusual scheduling.

EXAM À LA CARTE, 1978

A student threw a red checkered tablecloth over his exam table; set out three bottles of wine, corkscrew, glass, a plate of bread and cheese, and regulation No. 2 pencils before settling in for the test.

REVERSAL, 1982

Industrious hackers reversed all 199 seats in the 2–190 lecture hall so that they faced the back of the room. The prank was all the more ambitious because the seats are bolted to the floor.

PAPER AIRPLANE ASSIGNMENT, 1985

Students picked up the usual stack of handouts as they entered the lecture hall, only to find that one was a hack. The sheet gave detailed instructions for making a paper airplane and when to launch it at the lecturer.

TURBOJET, 1987

Hackers moved the massive turbojet on display in another building to the front of AeroAstro's unified engineering class. On the lecture hall blackboard, they asked, "Can You Say Turbojet?"

Reversal of all of the seats in 2–190, 1982.

GIANT CLACK BALLS, 1988

One morning, students and faculty arrived in one of the principal campus lecture halls to find an enormous collision ball apparatus suspended at the front of the classroom.

CHALKBOARD GREMLINS, 1981 AND 1992

Using a handmade radio-controlled device, a hacker raised and lowered the lecture hall chalkboards to the frustration of the lecturer in the 10–250 lecture hall.

6.001 SPELLBOOK, 1992

In a spoof on fantasy role-playing incantations, hackers distributed a "spellbook" to students that explained why some of their computer programming projects may go awry.

Large version of collision balls in a lecture hall, 1988.

WORTH A THOUSAND WORDS: HACKS FIT TO PRINT

Literary hacking at MIT can be traced all the way back to 1939, when an alumnus from the class of 1889 penned the novelty novel "Gadsby," a 50,110-word saga that did not contain a single "e." In fact, the author tied the "e" key down on his typewriter to avoid any inadvertent slips.

According to *Gadsby*, "Youth cannot stay for long in a condition of inactivity," and thus it is with literary hackers. Throughout MIT's history, hackers have been prolific authors, publishing parody magazines, posters, and brochures. The one thing they all have in common is obsessive attention to detail—each spoof copies the graphics and materials of its subject to the letter. Often, these hack pieces are mistaken for the original before the reader realizes, for example, that the official currency of the "People's Republic of Cambridge" is probably not changing to the ruble.

Hack publications were especially prevalent during the 1990s. The student newspaper *The Tech* was spoofed on many occasions including: *The Absolut Tech, The TeX Files, The Wreck,* and *The Dreck.* The Institute's official newspaper *Tech Talk* also was lampooned many times over the years, often with the sporadically annual *Tich Tolk.*

A variation on the printed hack genre is the written hoax, a good example of which is Professor R. Catesbiana's regular letters to the editor in the 1970s. This prolific, opinionated, but entirely fictional MIT professor made a point of taking a bizarrely contrary position on random issues of the day. For example, Boston residents were greatly alarmed by the frequency with which windows suddenly began falling out of the famous John Hancock building designed by architect and MIT alumnus I. M. Pei. So many popped out of their frames that a sizable portion of the building's blue reflective windows were soon replaced with plywood sheets. In a letter to the *Boston Globe*, Professor R. Catesbiana peversely complained when new windows replaced sheets of plywood. "It was distinct," he lamented. "Now, alas, it is just another skyscraper." The invention of MIT hackers, Rana Catesbiana (the scientific name of the American bullfrog) was assigned an office in the MIT lab where bullfrogs were kept for research on amphibian immune systems.

And then there's *VooDoo*, the sporadic campus humor magazine that's been spoofing life on campus and off since 1919. The boundaries of *VooDoo* humor have always reached well beyond the pages of its magazines. *VooDoo* stunts over the years have included a variety of elaborate hoaxes and publicity stunts, including a wild spree with a gorilla in a helicopter.

THAT VOODOO THAT YOU DO

VooDoo, "MIT's only [sic] intentionally humorous campus publication since 1919," has drifted in and out of print over the years. Especially in the period just before and after World War II, the magazine and its staffers played an important role in preserving the nascent tradition of anonymous campus pranks. The publicity stunts *VooDoo* pulled to sell magazines do not, by strictest definition, qualify as hacks, but they were born of the Institute hacking culture and they influenced it. A sampling:

WELLESLEY COLLEGE COUP, 1947

Listeners enjoying a Beethoven piano sonata on the Wellesley College radio station were perplexed when the music abruptly halted and was replaced by the ominous sounds of a studio ruckus. Suddenly, a male voice announced plans to set up "an electromorphic flux and contrapolar micro-reflector with which to create neutral mesons to blow up the college." His speech was followed by a commercial for *VooDoo* magazine. The incident did not cause quite as much of a stir as the event it spoofed—Orson Welles's "War of the Worlds"— but the press soon arrived to photograph the *VooDoo* staffers and the good-natured Wellesley announcers they had tied to chairs.

DIPSY DUCK, 1948

The magazine constructed a 12-foot-long "hydroelectric" Dipsy Duck, reporting it to be a prototype for the 250-foot-long Dipsy Ducks that would soon be installed beside rivers to generate electricity. The jig was up, however, when a motor was found to be powering the Dipsy prototype.

PSEUDO SCIENTIFIC AMERICAN, 1958

To celebrate the inaugural issue of *Pseudo Scientific American* magazine, *VooDoo* constructed a time machine. The staff members announced during the "test drive" of the gadget that they have decided to bring back the first Boston city planner. Boston history fans smiled when they learned that at the end of the stunt, a spotted calf appeared—a play on the conventional wisdom that most of Boston's winding streets are paved cow paths.

VOODOO GOES APE, 1965

Drawn by the anticipation of a good contest, crowds gathered in Killian Court to see the first annual American Faunch Championship—sponsored by *VooDoo*. The magazine claimed that the game dated back to the ancient Aztecs and is similar to soccer, with the primary exception of human goal posts. As police contained a burgeoning crowd, the game began but it soon halted as a helicopter descended on the playing field. It hovered over the ground briefly. Then a "gorilla" leaped out and tried to buy a copy of *VooDoo* from one of the players. When he learned the price has increased by a nickel to forty cents, a melee ensued. In the confusion, the gorilla grabbed the issue, pummeled several players, and departed the same way he arrived. The helicopter disappeared over the Great Dome and peace reigned once again on the playing field.

MIT students pour a beer to set the *VooDoo* hydroelectric Dipsy Duck in motion, 1948.

TEACHING A NATION TO MAKE SNOW: HOAX HACKS

At a school where gorillas leap out of helicopters to pick up a copy of the campus humor magazine, successful pranks are often elaborate affairs—no more so than the hoax hacks.

If the 1998 MIT Homepage hoax had not happened on April Fool's Day, hackers might well have tricked a few people into thinking that a Fortune 500 company had bought the world-class university. In response to MIT's increasing relationships with Disney, hackers were inspired to break into the web server and "Disnefy" the Institute's homepage. As a result, Mickey Mouse, in his Sorcerer's Apprentice costume, appeared to have conjured up a pair of mouse ears for the Great Dome. The banner headline, "Disney to Acquire MIT for 6.9 Billion," linked to a press release that announced: "Top-Ranked Engineering School Will Switch to Imagineering."

The MIT administration quickly responded with its own press release, "'Disney Buys MIT Hack Revealed by Low Price ..." In the release, MIT spokesman Ken Campbell noted that the Mickey Mouse Club theme song had long been a tradition at MIT, with the substituted lyrics as "MIT ... PhD ... M, O, N, E, Y."

The Baker House Snow hack began innocently enough in January 1968, when students from the Baker House dormitory reported a new engineering breakthrough to the press. If the environmental conditions were just right, the students claimed, they could make snow by opening the windows in their bathrooms and simultaneously turning on the showers full force. The *Boston Herald-Traveler* ran the story and photo on the front page. From there, the wire services picked it up, and before long the story swept the nation. Unfortunately for the media and those who tried to duplicate the experiment in homes and dorms across America, the MIT student announcement was a hack.

As hacks go, this one was a bit of an accident. Students had brought snow into the showers to make a snowman. Finding the powdery snow to be too dry, they turned on the showers in hopes of creating steam to make the snow more suitable for building a snowman. When the steam quickly filled the room, they opened the windows. The resulting foggy, slushy, snow scene struck them as so phantasmagorical that they felt it warranted a photo. The photograph was so interesting that it inspired the students to create a hack.

For their part, the hackers were surprised at the gullibility of the press. "Anybody knows that if you mix hot water and cold air," said one student, "the only thing you'll get is a cold shower."

A little more than twenty years later, in 1989 a small group of MIT hackers undertook a high-risk, high-profile hack with the aim of shaking the credibility of tabloid TV. At a student broadcasting conference, one of the hackers approached the executive producer of the *Morton Downey Jr. Show*, posing as a Harvard student and member of the North American Man-Boy Love Association (NAMBLA). The producer, thinking the topic would be ideal for ratings sweep week, took the bait and asked for literature about the organization. The hacker then requested materials from the NAMBLA organization and forwarded them on.

On the night of the taping, the hacker was seated next to Dr. Joyce Brothers as a chain-smoking Downey pummeled him with questions and insults, at one point tearing the NAMBLA brochure to bits. The audience was equally hostile, with one man shouting, "What they did to Ted Bundy, I'd do the same thing to you!" After the syndicated show aired, the hacker revealed the ruse to the media. In an interview in a local newspaper, he said, "My next step is to get on Geraldo, but I'm probably blacklisted from trash TV."

The story reveals something of MIT students' blind spots as well, however. Sometimes what is intensely funny in one context is less so in another. There was a fundamental naiveté that allowed the hackers to overlook the painful reality of pedophilia. It has always been true that the boundaries that define what is acceptable and appropriate for hacking—especially for off-campus activities—are difficult to discern. In the 1920s, it was permissible to drag a car up the side of a building but not to turn on the firehose and spray down the residents' rooms. Painting Smoot marks on the Harvard Bridge in 1958 made the participants nervous, enough so that they ran away when a police car drove by. Fifty years later, the police use the marks to locate accidents. A quarter-century ago, when a balloon suddenly emerged from the field of Harvard stadium during a football game, the crowd cheered. Today, in a post 9/11 world, it is likely the stadium would be evacuated. When is a hack not a hack anymore? No one can know with certainty. What is certain is that the next generation of hackers will continue to push the limits.

"PLEASE WAIT TO BE SERVED": THE PERFORMANCE HACKS

Sometimes a propeller beanie set jauntily atop the Great Dome says it all, but every now and then, a procession, a kidnapping, or even a boiling cauldron is necessary. In the vernacular, these "happenings" are known as "performance hacks." From mock swordfights to chanting monks, performance hacks are an enduring Institute tradition.

In 1978, two MIT traditions were melded to create one of the most beloved performance hacks of all time. Since 1953, the Institute had been crowning the Ugliest Man on Campus (today, it is the ugliest "manifestation"). The UMOC (pronounced "you mock") is a charity event held by MIT's service fraternity, Alpha Phi Omega. The man who collects the most money for charity is declared ugliest of them all. In 1978, the winner was also named Homecoming Queen for MIT's first home football game (in modern times). The spectacularly ugly UMOC, holding his cane aloft, rode into the stadium on a gargantuan float—a replica of MIT, complete with its own somewhat-misshapen Great Dome. The 1978 UMOC's bid to become national homecoming queen was thwarted when event organizers refused to allow him to enter the contest.

In 1996, the Order of Random Knights hacking group decided that Halloween was the ideal time for a performance hack. They made a goodwill tour of the campus, picking up the dead. Inspired by a scene in *Monty Python and the Holy Grail*, the hackers donned black robes and pushed wheeled carts piled with bodies around campus—the bodies, of course, were fellow hackers.

Hackers sometimes pull off a performance hack to help orient a new class of students to life at the Institute. Tradition has it that MIT upperclassmen take groups of first-years out for dinner on their first evening on campus to introduce them to the local eateries. They hold signs aloft with their intended destination—"Sushi Express," for example, or "TGI Fridays." During orientation in 1993, "Cannibals at MIT" offered frosh an alternative dining experience.

A "cannibal" in a chef's hat stirred an enormous cauldron while a table nearby was set for dinner. "Please wait to be served," patrons were instructed. Towering over the entire tableau was an enormous sign: *Cannibals at MIT ... We would like to have you for dinner!*

Wearing club T-shirts, the cannibals darted through the crowd measuring freshmen in search of the "perfect specimen." Those who passed muster were carried over and stuffed into the pot. When the pot was full, the Cannibals danced around it:

Cannibal chef prepares to cook up an MIT first-year, 1993.

Frosh be nimble, frosh be quick,
Cook frosh up on a great big stick!
and
Double, double, toil and trouble,
Fire burn and freshmen bubble!

In 1997, inspired by the movie *2001: A Space Odyssey*, hackers dressed as big, black, hairy apes and carried a massive black monolith into Killian Court during the frosh picnic. Apparently, their aim was to encourage the Class of 2001 to evolve into higher life forms during their years at MIT.

When the opportunity presents itself, hackers also like to orient parents to the MIT culture. During Parents' Weekend in 1991, they chose an appropriate venue—the official lecture for first-years and their families by Institute hack archivist Brian Leibowitz. The lights went down and Leibowitz began his slide talk.

Suddenly, ushered in by a loud, booming rumble, a contingent of commandos, dressed entirely in black, invaded the 10–250 lecture hall and grabbed a seemingly unsuspecting student. The commandos claimed to be representing SPODSA, the Secret Police of the Office of the Dean for Student Affairs. This was a clear slam at the ODSA for what students considered to be policies giving them less freedom. Two of the commandos read an arrest warrant detailing the first-year's nerdly crimes, which included caffeine abuse, accepting financial aid, and attending a lecture on hacking.

In a scene reminiscent of the abduction of Joe Buttle in the movie *Brazil*, the "unsuspecting" student was stuffed into a body bag and carried away. Leibowitz was then asked to sign a receipt, sign for his copy of the receipt, and sign for the commando's copy of his copy of the receipt.

A few months later, Leibowitz gave the talk again. As the event progressed, however, Leibowitz became aware that the audience was becoming distracted. What he could not hear was that the pitch of his voice was slowly—but noticeably—getting higher. About 15–20 minutes into the talk, his pitch was a half-tone above normal. As it turned out, hackers were controlling the audio output via equipment they had patched into the lecture hall sound system, running the microphone input signal through a mixing board that included a set of pitch shifters.

Leibowitz could not hear the audio feed over the sound of his own voice. Puzzled by the laughter preceding every punch line, it was only after the talk was over that he realized the sound had been distorted. In the meantime, the sound technicians running the lecture series had discovered the prank and fixed the problem. But the hackers weren't ready to call it a night. Shortly afterward, an unseen telephone began to ring on stage. The speaker searched for the source of the din and finally located it hidden in a nearby lectern. When he answered the phone, the hacker on the other end ordered a pizza for delivery to 10–1000 (the unofficial room number of the Great Dome). Leibowitz replied curtly that the location was outside of his delivery range and continued his talk without further interruption. The hackers had intended for the audience to hear the other voice on the line but when the technician disconnected the pitch shifter, he had unknowingly disconnected the feed to send the telephone call audio to the sound system.

Occasionally, performance hacks are directed at an audience outside the community, as in 1976 when hackers besieged a bus full of tourists making the usual trip down Amherst Alley to view the "natives." Much to the amusement of the guests, hackers attacked the bus with suction-cup arrows, Star Trek phasers, and other weapons. Students maintained that tour bus traffic doubled after the incident.

Baker House hackers attack tourist bus, 1976.

WHEN MIT WON THE HARVARD–YALE GAME: HACKING HARVARD

Over the years, the red brick school at the other end of Massachusetts Avenue has been the recipient of more than its fair share of hacks. Because it has been hacked so many times—in so many ways—Harvard may be the expert on MIT hacks, and the school has its own lore about the way it has been hacked by MIT. This account is written from the MIT perspective, however.

MIT hackers are particularly dedicated to "enhancing" the Ivy League school's most hallowed traditions. The earliest recorded MIT hack on "Hahvaad" was in 1940, when MIT students initiated a series of kidnappings of celebrity guests headed for Harvard. Members of MIT's Delta Kappa Epsilon (DKE or Deke) fraternity, posing as Harvard students, met Eddie Anderson at the Providence airport. The hackers told Anderson, who played Rochester on Jack Benny's radio show, that they were avoiding the crowds at the Boston airport and proceeded to take him to an informal get-together called a "smoker" at the DKE house at MIT. After a while, the hackers explained the hoax and transported him to Harvard.

A retaliatory "war" between the two schools ensued over the next two nights with MIT students claiming as their spoils 22.5 pairs of trousers, a belt, and a pair of undershorts. The next night Harvard thought they'd foil any further kidnapping attempts by setting a car and driver out in front of the Metropolitan Theater to transport Anderson to a reception. Alas, the DKE brothers went backstage, escorted him to a car waiting at another entrance and drove him to his scheduled event, infuriating the Harvard students.

The next year, an MIT hacker posing as a Harvard student picked up burlesque queen Sally Rand after a performance in the Latin Quarter. She was scheduled to entertain at the Harvard freshman smoker but instead was whisked to MIT where she was honored with a reception and the title, "Associate Professor of Entertainment Engineering." Rand was eventually delivered to Harvard and she noted that it was the nicest kidnapping she could ever remember. Later that same evening, hackers escorted the popular French chanteuse Yvette to MIT's Chi Phi fraternity house. After serenading the fraternity, she was returned to her rightful engagement at Harvard.

In 1990, Harvard launched a new annual tradition, and MIT was there to celebrate. According to the new custom, an 11-by-20-inch silver platter called the "Yard Plate," would be hidden in Harvard Yard for the incoming class to locate. Alas, the MIT Mole Hole hacking group found the plate first and delivered it to MIT President Paul Gray's office. Like a baby left on a doorstep, it came complete with note:

Harvard's Yard Plate is returned by President Paul Gray, 1990.

We give you this small token in appreciation
of your years of dedication to our traditional Tech values.
We hope you enjoy having this little bit of their tradition.
It might amuse you to know that the Harvard Class of '94
spends an evening searching Harvard Yard
for what is now in your hands …
Please feel free to dispose of the gift as you see fit.

President Gray promptly assembled a delegation of Institute luminaries and, in full academic regalia, they returned the icon to Harvard so that the new tradition could launch on schedule.

Over the years, MIT hackers have been especially keen on hacking the venerable Harvard–Yale football game. In 1940, they burned the letters "MIT" into the Crimson turf. Harvard lost that game and the team considered the incident something of a jinx. Ever since, Crimson officials have been on the lookout for MIT hackers.

In 1948, eight MIT hackers planted primer cord, an explosive used to ignite dynamite, under the turf of the Harvard football field. The plan was to again burn "MIT" into the turf. A groundskeeper discovered the ends of the wire hidden under the grandstand, however, and removed the cord but left the exposed ends in place to trap the hackers.

On the day of the big game, Harvard authorities spotted a student lurking near the wires attempting to conceal dry cells under his jacket. Apprehended, the hacker explained that "all Tech men carry batteries for emergencies." In tribute, many MIT students wore batteries under their jackets throughout the next week. Boston newspapers reported that the explosives would have blown a crater in the field, but tests performed during the planning stages of the hack showed that the primer cord would have left only a shallow turf burn.

In 1978, just a few days before the big Harvard–Yale game, several Yale students were caught in the act of burning a large "Y" into the field. While repairing the turf, the Harvard grounds crew discovered a remote control spray-painting mechanism under the field. If triggered, the device would have painted "MIT" on the field but, again, the attempt was foiled.

It wasn't until 1982 that MIT was again successful at hacking the Harvard–Yale game, pulling off multiple legendary hacks in one afternoon. Harvard had not scored against Yale since 1979, but now, with 7:45 minutes remaining in the second period, it had just chalked up its second touchdown. The atmosphere in the stadium was charged. Suddenly a weather balloon burst from the turf at the 46-yard line. As it inflated to six feet, the spectators in Harvard Stadium could make out the oversized white letters that spelled "MIT" before the black balloon burst in a cloud of talcum powder. This hack was the first of three MIT plays.

Wearing the signature white pants and sweatshirts of the Yale Marching Band, the forty or so students comprising the MIT Band made it past security guards before the start of the game. At half time, they paraded onto the playing field, stretched out in careful formation and, with their prone bodies, spelled out "MIT."

EXAMINING BLAST BATTERIES—Patrolman Peter A. Coletta and Harold J. Finan at Station 14.

Police officers examine blast batteries from an aborted hack, 1948.

GLOBE ARTIST'S SKETCH shows how M. I. T. letters wired with dynamite caps were placed on 50-yard line at Harvard Stadium.

A newspaper sketch of an aborted hack, 1948.

Almost fully inflated MIT Balloon at the 1982 Harvard–Yale football game.

Bob Brooks captured the sequence of the Harvard–Yale balloon hack from beginning to end.

Cementing their win, hackers passed out cards in the bleachers during the last quarter, telling Harvard fans that, en masse, they would spell "Beat Yale!" On cue, all 1,134 spectators raised their cards—not knowing that they spelled "MIT." Few people who attended that game could tell you today whether Harvard or Yale had scored more touchdowns, but they have never forgotten the team that "won" the game. "I thought it was fabulous," Los Angeles Dodger first baseman Steve Garvey, who happened to be in the stands, told the *Boston Herald-American.* "We never had anything like that at the Michigan–Michigan State game."

While this hacking hat-trick was definitely a coup for MIT, the three hacks were not a coordinated effort but instead the work of three independent hacking groups that had coincidentally set their sights on the same target.

In 1990, Harvard campus police were patrolling the stadium in the days leading up to the Harvard–Yale game but this did not deter hackers from MIT's Zeta Beta Tau fraternity. Their aim was to send a rocket-powered banner over the Harvard goal post during the big game, and a little police presence would not stop them.

With exceptional perseverance, the hackers soldiered on, even after Harvard police discovered and removed the apparatus, which had been concealed under sod at the goal line. From that point on, a row of police cars kept a vigil, their headlights trained on the field after dark. The hackers timed their reinstallation between police shifts, taking care not to leave footprints. This time, they used butter knives to bury the wire that ran from the rocket, under the field, and into the bleachers.

Because they could not test the apparatus on the field, they devised a back-up method. They placed a multimeter between metal bleachers to test the circuit, then checked for resistance between the two bleachers. If the multimeter read 3 ohms, they knew the hack would go according to plan.

On game day, one of the hackers connected the wires to the battery pack he was wearing in the inner pocket of his jacket, then launched the model rocket engine across the playing field and over the goal post, trailing a banner that read "MIT." As they had in 1982, local newspapers proclaimed MIT the winner of the game, and the Zeta Beta Tau hackers had the satisfaction of demonstrating just why "all Tech men carry batteries."

Not content to dominate the playing field, hackers in 1982 actually made an attempt to take over Harvard—peacefully and bureaucratically, of course. MIT student government leaders had been reading that Harvard's student government was in the throes of reorganization. The MIT regime saw this as an excellent opportunity for a hack. They passed a resolution granting

Harvard colonial status and appointed an MIT student colonial governor. (Even Harvard students got in on the fun as seen from an article in the Harvard *Crimson*: see "Student Leaders at MIT Claim Harvard as Colony.") The prank continued into the following academic year. During MIT Undergraduate Association President Kenneth Segel's speech at the first-years' picnic for the annual Rush/Orientation, the MIT hacking group Commando Hacks (pretending to be "Freedom Fighters" from the Harvard Colony) staged Segel's abduction. Other members of the hacking team unfurled the banner "Free Harvard," and, finally, a ransom note was presented to MIT President Paul Gray demanding that Harvard be released from the "tyrannical rule of the engineers." (Segel would be released without explanation, and Harvard was a colony no more!)

MIT Undergraduate Association president's abduction to free Harvard from MIT colonization, 1982.

STUDENT LEADERS AT MIT CLAIM HARVARD AS COLONY

Jessica Marshall

MIT's colonization of Harvard was described in detail by the "colonists" themselves in the April 21, 1982, edition of the Harvard Crimson.[1]

Anyone who wonders just where to find the center of student authority at Harvard can now look to MIT. Confronted with rumors that Harvard currently lacks a cohesive student government, the MIT Undergraduate Association last Thursday passed a resolution granting Harvard College colonial status and appointing sophomore Paula J. Van Lare colonial governor.

Van Lare said yesterday she does not expect any organized resistance from her new subjects. "Seeing the success Harvard has had organizing itself in the past, I sort of doubt it," she said.

MIT junior Kenneth H. Segel, president of the MIT student government, said yesterday giving Harvard colony status will be no problem, explaining that "basically everything can be reduced to an engineering problem." He added, "We'll probably let you move up to province status if you're good."

[Harvard] Dean of Students Archie C. Epps yesterday expressed surprise at the MIT takeover, saying, "I didn't realize that they learned anything about American government at MIT."

And Associate Professor of History Bradford Lee questioned the prudence of the move, observing that the invocation of colonial status "usually breeds an anticolonial movement."

Confirming Lee's prediction, Andrew B. Herrmann ('82)—former chairman of Harvard's Student Assembly—called for a "26-mile blockade around Harvard Square." Asked Herrmann, "What are they going to attack us with—calculators and slide rules? They don't even have a football team."

Harvard's new rulers include the leaders of the MIT Undergraduate Assembly elected in early March. Segel and his running mate, vice-president Kenneth J. Meltsner, also a junior, ran on the Gumby Party, whose motto is "Reason As a Last Resort." After their election, they found that the MIT student body "pretty much had their stuff together" and turned their sights to Harvard.

Besides governor Van Lare, the protectorate also includes sophomore William B. Coney, as secretary of defense, and Deren Hansen, as ambassador to extraterrestrial civilizations. Van Lare said she expects to create other posts for her MIT friends.

Harvard's current Student Assembly chairman Natasha Pearl ('83), when informed that her powers had been usurped, dismissed MIT's gesture of munificence, saying that her own unfunded government doesn't "want their sympathy, just their cash or personal checks."

1 © 1982 *The Havard Crimson*, Inc. All rights reserved. Reprinted with permission.

In 1995, one of the staff members from MIT Facilities decided to play a small prank on his coworkers in retaliation for all the ribbing he had gotten for being a Harvard graduate. That spring, a section of sidewalk at the corner of Ames and Main streets was replaced following utility work for the construction of Building 68. Imagine his fun showing his colleagues a brass "H" embedded in the concrete. Not to be outdone, several Facilities members worked with staff from the machine shop to fabricate a brass H being gnawed by a beaver. This artifact was embedded in the sidewalk in front of Building 10 in Killian Court. All MIT students entering and graduates exiting the campus at graduation would have to "walk all over Harvard."

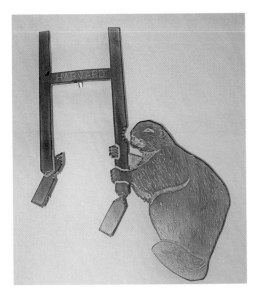

Brass beaver gnawing on the Harvard football goalpost, 1995.

The hack was removed swiftly per orders from the administration in order to discourage further hacks in Killian Court. The "H" was also only on display for a limited time because during the summer of 1995 the City of Cambridge had to repair a water main break at the intersection. Facilities colleagues tried to reach the site before it was destroyed by jackhammers but were unsuccessful.

Finally, no survey of Harvard hacks is complete without discussion of the Harvard Bridge. Nicknamed the Mass. Ave. Bridge, the Harvard Bridge is a primary vehicular and pedestrian thoroughfare connecting Boston and Cambridge. Back Bay sits on the Boston side, MIT on the Cambridge side. Harvard's main campus is located several miles farther down Massachusetts Avenue on the Cambridge side.

The reason for this anomaly is history. When the Harvard Bridge was dedicated in 1891, MIT was not located on the bank along the Charles River but in the middle of Back Bay in Boston. The bridge-naming controversy began in 1916 as soon as the MIT campus moved to Cambridge, and it has continued ever after. On several occasions, when the bridge has been closed for renovations, proposals have been submitted to rename the bridge for MIT. Many at the Institute have considered the span an embarrassing example of civil engineering and have not wanted MIT associated with it. They, at least, have been pleased when name-change attempts failed to make it through bureaucratic channels.

Hackers, on the other hand, have always felt the Harvard Bridge moniker was a wrong that needed to be righted. They have hacked the bridge a half-dozen times and left MIT's indelible mark with Smoots. In 1949, when the bridge was closed for renovations, they changed the "Harvard Bridge Closed" sign to "Technology Bridge Closed." When the bridge opened later that year, the governor of Massachusetts was about to lead a motorcade over it when a convertible cut in front, replacing the official car at the head of the procession. The renegade car carried two clowns, a brass band, and ten members of *The Tech* newspaper staff wielding a sign that read, "The Tech Dedicates Technology Bridge." State police intervened at the last moment, unfortunately, and waved the governor's car ahead, so the *Tech* car was ultimately the second to cross the bridge.

MIT students attempt to be the first to cross the newly renovated Harvard Bridge. Hackers had replaced the original construction signs renaming the structure the "Technology Bridge," 1949.

THE JOHN HARVARD HACKS

MIT hackers long have had a penchant for pranks on Harvard icons and traditions. The bronze statue of the college's founding father, John Harvard, has proved an irresistible target over the years. Designed, ironically, by MIT alumnus Daniel Chester French, the statue has presided over Harvard Yard for more than a century.

JOHN HARVARD'S PORTO-POTTY SERVICE, 1960

Hackers positioned a toilet stall door in front of the statue along with the advertisement, "Johnny on-the-Spot, Portable Toilets, Rented/Serviced, Boston, Mass. Capital 7–0777."

JOHN HARVARD'S BRASS RAT, 1979

John Harvard proudly sported a Class of 1980 Brass Rat, the MIT school ring. To avoid damaging the statue, hackers cast the bronze ring in two parts scaled to the statue's proportions and then use epoxy to join the two halves on the statue's finger.

John Harvard sports MIT's famous class ring, the Brass Rat, 1979.

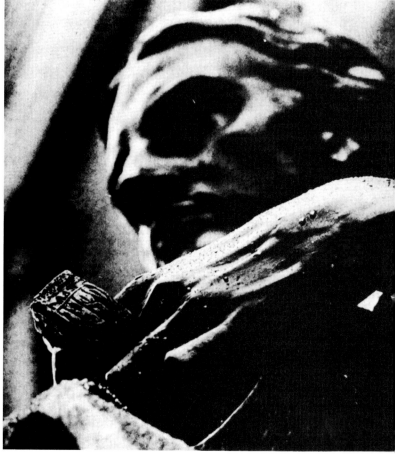

Close up views of the giant
Brass Rat before and after it
was attached to John Harvard's
finger.

JOHN HARVARD'S BAD LEG, 1990

To underline the trouncing that both MIT and Yale had just given Harvard at the Harvard–Yale game, MIT hackers placed a cast on the statue's right leg, a skull cap on its head, and a regulation MIT Medical Department brace around its neck accessorized with a pin that read "Ask Me about My Lobotomy."

JOHN HARVARD'S VISIT FROM SANTA, 1996

This John Harvard hack actually took place at MIT. Hackers placed a sack of coal at the bottom of a stairwell. Positioned above it was the red-suited arm of Santa Claus, who had apparently just dropped the bag. The sack was addressed, "To John Harvard. From: Santa Claus. You've been a baaaad boy, Johnny!!"

JOHN HARVARD, UNIBOMBER, 1996

When it turned out that Unibomber Ted Kaczynski was a 1962 Harvard graduate, MIT hackers dressed up John Harvard to fit the FBI mug shot of Kaczynski distributed in the media—complete with white hooded sweatshirt, unruly hair, moustache, and sunglasses. He also sported a pair of handcuffs.

BECAUSE IT'S THERE:
THE BEST OF THE REST

Often, when a prominent fixture emerges on the campus landscape, members of the MIT community will say, "Well, it's just a matter of time." And everybody knows what that means. The statue, the sign, the vacant lot is just asking to be hacked.

When a new information booth, The Source, was set up in the student center in the mid-1990s, hackers could not resist setting up a spoof booth a few yards away. The sign for The Sink looked nearly identical to that of the neighboring booth, but the personnel staffing The Sink dressed entirely in black, wore dark glasses, and looked furtively about them as they attempted to collect information from passers-by. (The joke derives from basic electrical engineering terms describing local electrical currents.)

MIT information booth, aka The Source in the Stratton Student Center, before being hacked, 1996.

Elevators at MIT are another obvious target, although usually they are hacked for utilitarian purposes—to reach a forbidden floor, for example. In fall 2000, however, hackers transformed both sets of elevators in the student center from mindless machines to erudite and sometimes sarcastic "cybervators." When a rider pressed a floor number, the cybervator would give a summary of what might be found there, sometimes with biting commentary. Those pressing the basement button might hear a message saying, "Video arcade, post office, bowling alley," or "There's a basement? Who knew?" If they chose floor 1, the elevator might warn, "expensive groceries, expensive banking, expensive clothing, expensive food." On 2: "black leather, twisted art, and pango-pango wood." In all there were twelve messages. Rumor has it that when a campus police officer entered an elevator to investigate, he was warned off: "This is not the hack you're looking for."

The Source becomes The Sink, 1996.

Earlier that year, hackers were inspired by another utilitarian machine, the ice-cream vending machine in Lobby 16. When a customer selected an item, a large suction arm—similar to that on a vacuum cleaner—would select the product from the freezer and deposit it into the vending slot. This mechanism reminded hackers just a little too much of the aliens in the movie *Toy Story*, who worship the Claw that delivers them to waiting children. They placed little green Toy Story aliens inside the machine along with a sign spoofing lines from the movie:

The vacuum is our master!
The vacuum chooses who will go
and who will stay!

An ice cream vending machine is redesigned to pay tribute to *Toy Story* and its Claw, 2000.

In fact, hackers have often employed animated characters in the service of their creations. When a series of renovations and additions left an inexplicable wall at the bottom of a stairwell in Building 9, hackers propelled the beleaguered Warner Brothers' critter Wile E. Coyote smack into the dead-end. The hack was actually two murals—one of Wile's outline as he tried to pass through the tunnel and another, on a side wall, of the empty box for his ineffectual Acme Instant Tunnel Kit. The murals still adorn the walls—one of the few hacks not to be removed—perhaps as a precaution to unsuspecting pedestrians.

Grumpy Fuzzball is a homegrown, fuzzy, heavy-lidded cartoon critter that hackers made appear at the height of final exams in 1989. Grumpy Fuzzball replaced the owl icon on signs for Athena, the campus network of computers, and also replaced the icon on the log-on screen. Athena staff determined the prank was harmless but still worked to repair and restore the original code in order to erase Grumpy. They were unsuccessful, but the icon spontaneously reverted back to his wise old self the next morning, just as hackers had planned.

In 1978, the hacking group James E. Tetazoo kidnapped the life-size cardboard cartoon Renaissance Man who served as the Independent Activities Period (IAP) mascot. (IAP is an interval between the fall and spring semesters when members of the MIT community can take short courses in everything from wine tasting to boat building.) The hackers left a photograph and ransom note at the scene. The photo showed the cartoon man bound, gagged, and being hauled off by a hooded kidnapper. A copy of the *Boston Globe* was propped in the upper right-hand corner of the frame to establish the date.

In their ransom note, hackers made a variety of demands, including that the drop date for courses be extended to sixteen weeks, that all Institute courses be taught in the computer programming language COBOL, and that light observe the 55-mph speed limit. As requested, administrators replied to the kidnappers' demands in *Tech Talk*, the official campus newspaper. In an open letter to the mascot's abductors, they agreed to many of the stipulations. Administrators thought, for example, that slowing down light might be a boon for physicists trying to study it. But they rejected two of the demands: "People can teach classes at MIT in whatever color they choose. As for the 16-week drop date, it simply doesn't go far enough." In the end, Renaissance Man was returned.

Squanch is another MIT cartoon figure. *HowToGAMIT*, a student-produced handbook to MIT culture and campus, defines "Squanch" as: "(1) resident of East Campus Third East. (2) A short fellow with a picket sign and a wilted flower." Mostly, Squanch is a logo, but the cartoon-like figure has appeared in one notable hack when James E. Tetazoo designed a garden in Squanch's

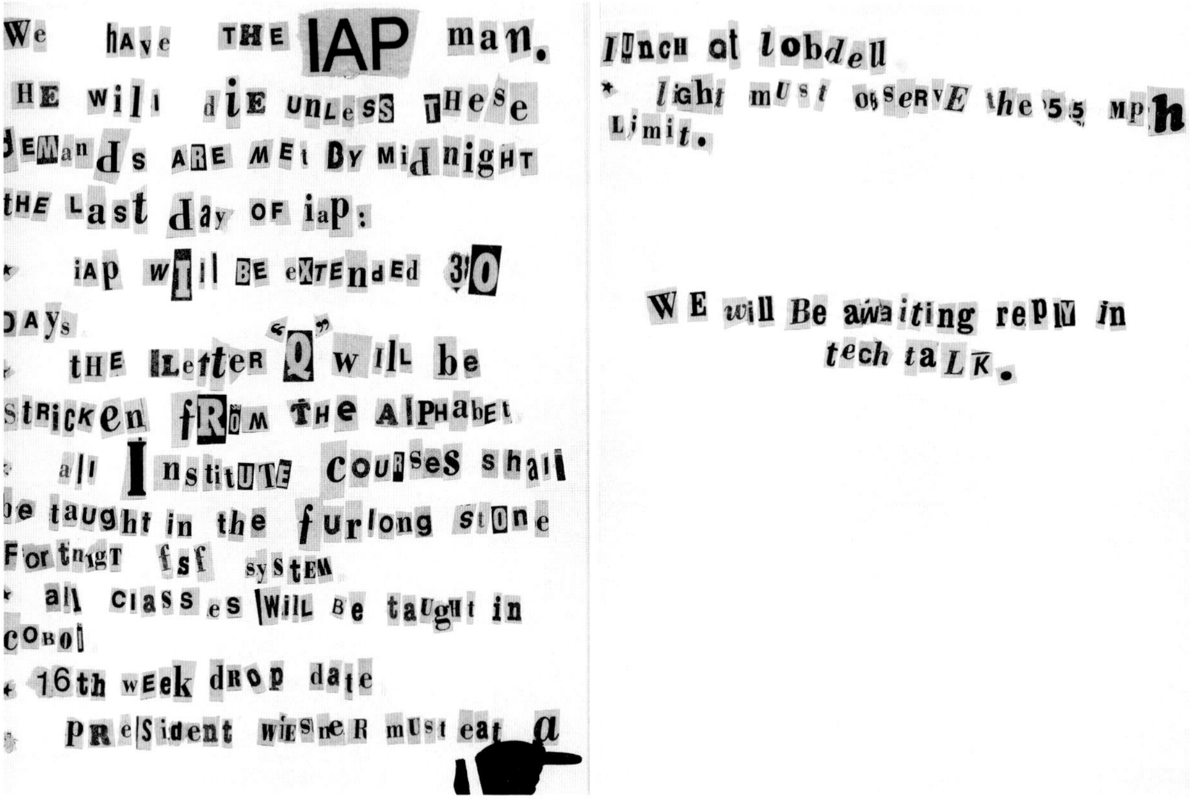

We have THE IAP man.
HE will die unless these
demands are met by midnight
the last day of iap:

* iap will be extended 30
days
* the letter "Q" will be
stricken from the alphabet
* all institute courses shall
be taught in the furlong stone
fortnigT fsf system
* all classes will be taught in
COBOL
* 16th week drop date
* president wiesner must eat a

lunch at lobdell
* light must observe the 55 mph
limit.

WE will Be awaiting rePly in
tech talk.

Ransom letter for the IAP Renaissance Man, 1978.

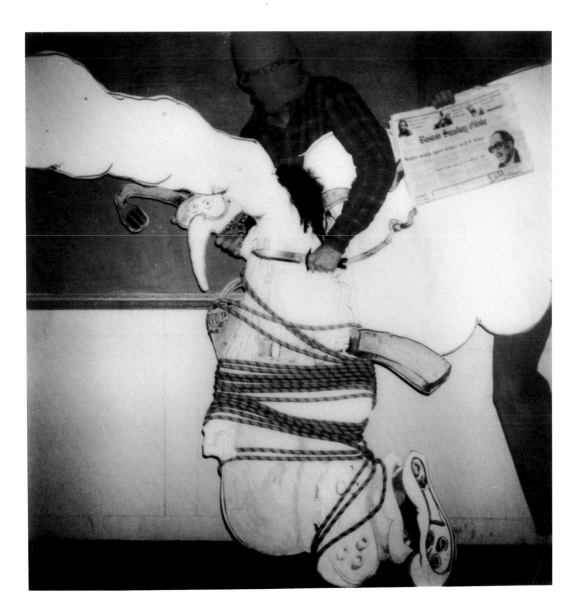

Kidnapping of IAP Renaissance Man, 1978.

image in December 2000. The physicists working in the buildings overlooking the Squanch garden, which is enclosed by Buildings 2, 4, and 6, decided they liked the new landscaping just fine, and it has never been disturbed.

After a century of hacking, the entire campus has become infected with the spirit. Staff and administrators get into the act from time to time. One legendary nonstudent hack was the 1982 Building Cozy hack. While perusing the book *Items from Our Catalog*, a spoof on the L. L. Bean catalog, the Institute's real estate officer, Philip Trussell, was struck with the hacking bug. He submitted a purchase order to Avon Books, publisher of the parody, for Item #6RMS RiVVW, a building cozy, to protect the 23-story Green Building. When the cozy did not arrive, there ensued a correspondence between the MIT hacker and the publisher of the fake catalog.

The employees of the Real Estate Office had been enjoying this joke between their boss and the mock catalog company when, a few weeks later, a large flatbed truck pulled up out front. Nearly everyone in the office was flabbergasted to discover that the truck was hauling a massive crate labeled "Building Cozy."

HACKING BY MAIL

Bitten with the hacking bug, real estate officer Philip A. Trussell attempts to order a "building cozy" from a parody mail-order catalog.

December 17, 1982

Gentlemen:

A month ago we forwarded to you our purchase order #ES437192 for a building cozy for our Earth Sciences building: color—red check.

We have not received confirmation of our order; however, if the shipment is in the mail (or possibly on a Conrail car), we will forward our check in the amount of $51,575.25 as soon as the order is received.

Thank you for your attention to this matter, and we wish you best wishes for a successful holiday season!

Sincerely,
Philip A. Trussell

January 10, 1983

Dear Mister Trussell:

We are in receipt of your order for Item #6RMS RiVVW, Building Cozy, Red Check model. Unfortunately, we have just received a similar order from the Port Authority of New York and New Jersey. Since this will undoubtedly consume the entire supply of good, honest American red-checked gingham that we might possibly import from Hong Kong over the next few years—to say nothing of the entire East Coast supply of seamstresses and seamsters—I am afraid that we will not be able to guarantee delivery of your order before the next ice age cometh.

In the meantime, perhaps you should consider individual orders of the Solar Watch Cap, which is sturdy enough to warm the most professorial brain.

Yours sincerely,

Ann C. McKeown
Publicity Manager

Building Cozy arrives . . . before next ice age, 1983.

ZEN AND THE ART OF HACKING

The essays in this section offer first-person perspectives on the art and science of hacking. The authors' philosophies and anecdotes illustrate why, generation after generation, regardless of changes in trends and mores, hacking thrives at MIT.

IT'S NOT A JOB, IT'S AN ADVENTURE
David Barber, 2003

As the Confined Space Program Coordinator at MIT, I and my partner Gary Cunha evaluate, manage, and remove hacks at the Institute. To the best of my knowledge, we're the only people at MIT—maybe the only two people anywhere—whose job descriptions include "hack management." This unusual duty has given us a special perspective on hacks and the people who perform them.

Whenever a hack appears on campus, Gary and I are asked to come up with a game plan for either allowing the hack to remain or for immediate removal. Here are the basic criteria we use to determine whether a hack should remain:

1. Safety to the public
2. Damage to any MIT property
3. Potential damage to the environment
4. Changing weather conditions
5. Resources necessary for removal
6. Whether the hack was "expected"

Sometimes the hackers contact me directly to inform me of their latest creation. These communications usually arrive overnight, often in the form of anonymous email messages that alert me to the installation of the hack, give its reasons for being, and request a specified stay of execution. Occasionally, the hackers' method of communicating with me can cause almost as much commotion as the hack itself. I recall once when the communication was left in a cardboard box by my office complex and someone mistook it for an incendiary device.

The hackers may also provide me with the details of how they put the hack together, a description of the precautions they have taken to protect the community and the property, and instructions on how to disassemble the hack in the safest and most efficient manner. In the midst of the removal process, we often find little surprises, an edible treat, for example, to provide us with a smile as we undertake our duties.

Perhaps our most famous—and certainly best attended—removal, with more than 300 people on hand, was the dismantling of the frieze in the lobby of Building 7. Hackers had "edited" the wording of the Institute credo to read "Established for Advancement and Development of Science its Application to Industry the Arts Entertainment and Hacking." This hack was the most ingenious we'd ever removed. The texture of the materials used, the match of the coloration to the existing inscription, the carving of the letters, and the placement of the frieze were of a caliber befitting the best traditions of hacking at MIT.

In fact, this hack was installed so skillfully that, unless you were looking for it, you probably would not have known it was there. It was in place for two days before anybody noticed it—a testament to the quality of the hackers' craftsmanship and their attention to detail. The hackers also exhibited elaborate care to ensure that they did not damage the installation's surroundings and graciously provided MIT's director of facilities Victoria Sirianni with an anonymous note detailing how the frieze was constructed and how it could be removed without damaging the building.

Communicating with the hacking community has been one of the more interesting facets of my job over the years. At one point, I was approached by a journalist who was interested in writing an article about hacks in general and MIT hacks in particular. The journalist and I had several conversations relative to the MIT tradition of hacks and the hacking community. This person then asked if it would be possible to get an interview with a current member of MIT's hacking community. I posed this question to some people I believed had access to the hackers. Eventually, I got a blind conversation going with a group of the hackers about this possibility. The journalist wanted to interview anonymously, stressing that no names would be used, and offered to provide aliases for the hackers who were interviewed. After several rounds of parallel conversations, it was decided that the journalist would ask me the questions via email and I would relay them to the hackers. The hackers would then send the answers back to me and I would forward the answers to the journalist, after removing all references to the hackers' identities from the electronic correspondence. The magazine got its article and the hackers' identities remained secret. In fact, I still don't know the identities of the people with whom I was in contact.

Inevitably, the hackers eventually decided to hack the hack removal team itself. My hacking response team, the Confined Space Rescue Team (CSRT), carried our rescue equipment in a converted ambulance. Whenever we were in the midst of "hack removal" activities, our vehicle would be parked nearby, so it came to be synonymous with the CSRT. One morning we discovered that our response vehicle had been transformed into a Hackbusters mobile. The hackers had adorned our converted ambulance with clear appliqués (1- to 2-feet tall, and several feet in length) identifying it as the Official Hackbuster vehicle in the style of the vehicle used in the *Ghostbusters* movies. "Hack" was circled in red and had a diagonal line crossed through it. True to the attention to detail that hackers are famous for, the Hackbusters hack came complete with a note to us that included instructions on how to best remove the decals. The note even informed us about the makeup of the materials used in the hack, explaining that they were safe and would not harm the paint on the vehicle.

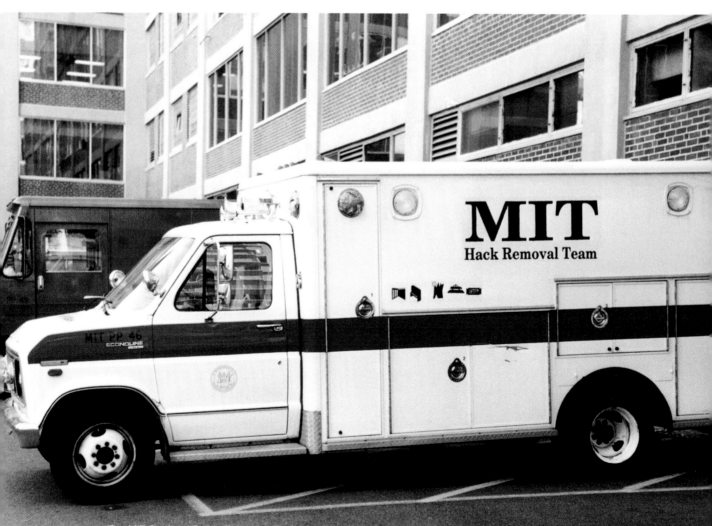

The MIT Confined Space Response Team vehicle turned into the MIT Hack Removal Team van.

Historically, the Great Dome has been the number-one target for hacks. Over the years, it has been adorned with a piano, a barbell weight, and of course the infamous police car. I have also seen it transformed into a variety of forms, including a giant beanie cap. On that occasion, hackers draped the dome with red dyed-silk parachute material cut into strips to create the visual effect of a striped beanie cap and topped it with a propeller with 28-foot-long blades that were so perfectly balanced, they rotated in the wind. Most recently, the dome sported the gold ring from *The Lord of the Rings*, complete with authentic inscription. This particular hack came complete with a "hack-back" guarantee: If we would leave the hack in place up through the opening date of the movie, then the hackers would remove the hack. We did and they did.

Some years hackers are more prolific than others. One year, for example, we had four hacks in the course of a single week: a tribute to Douglas Adams, the Magic Pi Ball on Building 54, a dinghy in the moat of the MIT Chapel, and the barbell weight on the top of the Great Dome.

As MIT's official hack removal team, we wait and wonder what will be placed on the Great Dome next or what remarkable transformation will occur somewhere else on campus. We also speculate about how hackers will be inspired by some of the new buildings under construction at MIT.

With the symbiotic relationship between MIT and the hackers, it seems only fitting that some space be found where historic hacks can have a permanent home, a place that is worthy of their contribution toward making MIT the most interesting place to work in the world.[*]

David M. Barber's current job title is Emergency and Business Continuity Planner for MIT's Security and Emergency Management Office. He gets the first official look at most hacks performed at MIT.

MASTERY OVER THE PHYSICAL WORLD
André DeHon, 2003

I've always enjoyed a good hack, so as a freshman entering MIT, I was delighted to get my first-choice housing selection. It meant I would be able to associate myself with a hall in an MIT dormitory that took as its motto "Hackito Ergo Sum." I hack, therefore I am.

As a graduate student, and now as an alum, I've made a point of photographing MIT hacks, collecting interesting tidbits, and helping share those with the world—in large part through the IHTFP Gallery, a website devoted to MIT hacks at <http://hacks.mit.edu>. From time to time

---

* Many historic hacks now can be found in the Student Street at the Stata Center.

people ask me why a busy alum bothers to spend time documenting hacks. The answer lies in understanding the relationship of hacking to MIT.

Hacking, while playful and irreverent on the surface, is really very much about MIT's value system. Hacking, perhaps more than anything else, gives the world a glimpse into the complex ecosystem that makes MIT such a special environment. To use another metaphor, it's like a shard of a hologram—it's not the whole, but it is a piece through which you can see what the whole might be like.

Hacks provide an opportunity to demonstrate creativity and know-how in mastering the physical world. In that respect, they reflect the Institute's value system. At MIT, intellect and its application are valued, as opposed to athletic prowess, for example. So, while other schools get excited about their sporting teams, MIT students eagerly anticipate the annual mechanical engineering design contest. And the next hack. MIT is not about "our jocks beating your jocks" but "our nerds mastering the physical world" by, for example, putting a campus police cruiser on top of a 9-story-tall dome or inflating a balloon in the middle of the field during a Harvard–Yale game. It's not "we can run faster than you," it's "we can manipulate the physical world to do things you hadn't imagined were possible." As character Chris Knight says in the 1985 film *Real Genius*, it's about "using your mind creatively."

An important component of many hacks is to help people see something in a different way, to give it a humorous, satirical, or poignant twist. As Eri Izawa notes in her "Engineering in Action" essay, giving people a jolt out of the intense atmosphere at MIT reminds them to smell the roses and helps put the day-to-day trials into a big-picture perspective. And, of course, the MIT administration occasionally does or proposes some pretty silly things, like putting a giant hairball outside the cafeteria in the student center. There was a sequence of at least three hacks surrounding the whole hairball controversy, and they helped focus some of the discussion at the time. Fortunately, there was still no hairball in front of the dining facility the last time I visited Cambridge!

Hacks are good show for the outside world, they are good morale boosters for the MIT community, and they really are good for the students involved—whether or not they realize it at the time. It encourages them to think creatively, an essential skill for innovative engineers and scientists, and it gives them experience solving real engineering problems.

MIT hackers thrive on new ways to view old venues (hmm, Lobby 7 would make a nice cathedral) or current events (we need to demonstrate the folly of a giant hairball in front of the cafeteria). They generate visions of the way the world might be and then turn them into reality—

with no artificial constraints, supervision by elders, or safety net. Hacking is an excellent training ground in the small for acquiring the skills and experience necessary to be visionaries and doers in the large. Hacks attack focused problems not circumscribed by grades or artificial rules, but by real-world engineering constraints (costs, difficulty, physical properties of materials, access, politics). They are not judged on effort or conformity but solely by success, impact, and personal satisfaction. These experiences are the key ingredients to nurture excellent engineers, managers, and skunk-work teams.

Yes, classes have their role. It is necessary to know core facts and analysis and get focused experience using this knowledge and these techniques (labs, projects, homework). Classes, however, must be tailored to the large volume of students. Problems must be of a narrow enough scope to be doable by everyone and must have considerable uniformity to be gradable by limited course staff. For this reason, hacks are a real complement to a student's formal classroom and lab training.

In hacking, there is no teaching assistant or professor to bail you out when things don't work. Just like in the real world, you have to find a way to solve the problem yourself. This forces you to tackle the whole problem, the whole experience, and it builds confidence and independence. MIT hackers learn to envision new things. They learn management, delegation, teamwork, planning, failure analysis, and public relations. As Izawa writes in her essay, they really get to see "engineering in action" in an open-enough setting to be real, and they see this on a timescale short enough to facilitate rapid learning. They learn that they can turn their own visions into reality, and they learn how to do it.

It's hard to imagine a more valuable tradition for an institution like MIT. The fact that this tradition was not handed down by the faculty or administration, but evolved among the students and continues to thrive, is indicative of the special environment that is MIT.

POSTSCRIPT, 2010

Since *Nightwork*, I've spent a decade on the other coast (seven years of which at That Other Institute of Technology) and am now a tenured faculty member in the Ivy League (Penn). I've participated in and led engineering research projects; I've seen pranks at other schools; I've been the subject of pranks (it's great to know that you inspire your students to action!); and I've watched MIT hacks unfold from the other side (I was on campus in Pasadena when the Fleming Cannon [Caltech Cannon to most of the rest of the world] disappeared). This perspective has only served to reinforce the position I articulated in this essay. MIT hacking has been and can continue to be a great asset to MIT students and the Institute.

André DeHon spent a decade at MIT (1986–1996) observing MIT hacking culture, during which time he also managed to finish several degrees in electrical engineering and computer science. He is now an associate professor of electrical and systems engineering at the University of Pennsylvania.

ENGINEERING IN ACTION
Eri Izawa, 2003

Hacking at MIT has nothing to do with the popular image of breaking into computer systems. Instead, it is the production of clever and benign pranks meant to amuse and awe the public; in their ideal form, hacks are a melding of art, inspiration, and engineering. Hacks range from the famous campus police cruiser on the Great Dome and the MIT-labeled weather balloon in the 1982 Harvard–Yale game, to less well-known pranks, like the giant slide rule in the Student Center atrium and the Christmas lights that spelled out "MIT" in 6-foot-tall letters on the Little Dome. Hacks have become a way of life at MIT; hardly a term goes by without strange objects appearing in odd places.

But as common as hacks have become, they still require a great deal of insight, hard work, and old-fashioned engineering. Although some hackers may not fully realize it at the time, the production of a successful hack follows the same creative and technical process as any large-scale engineering project—with the added complication that, along with traditional issues such as cost, safety, and manageability, hackers must avoid getting caught by the campus police! (Though many hacks happen in public places, some are set up on places such as rooftops, which are not meant for public access.)

The first step of planning any hack is to decide upon the final product. Often, this is the "What if?" stage. The giant Plexiglas® slide rule that appeared in the Student Center, for example, was the work of students who, tired of the plans for an unpopular shaman's hat sculpture made with human hair, decided to construct something more symbolic of MIT's "nerd image," while providing an unobstructed view of the atrium. Hence, they decided to build a slide rule made of transparent Plexiglas®. Of course, sometimes hackers just decide that it's time to "pull a hack" for no particular reason at all.

Hacks must also pass a "couth" (the opposite of "uncouth") test to be considered a true hack, as opposed to an obnoxious and worthless prank. Hacks must be amusing and generally benign. For example, an idea that may be interpreted as racist would be discarded. The making of fake notices of student loan cancellations, which resulted in hysterical phone calls to the Bursar's Office, is a case where the perpetrators failed to apply the couth test. In addition, hacks that cause permanent structural damage are avoided. Hackers believe that property damage is not only inappropriate but bad for future hacks, since repair costs could result in an administrative backlash against hackers.

Once the general idea has been established and accepted, the design begins. As with any engineering project, cost is an important initial factor. Funding comes exclusively from hackers' pockets. Since many hackers tend to be undergraduate students, as one anonymous hacker puts it, "they also tend to be poor." For a group of ten hackers, the cost of a hack rarely exceeds $100, or $10 per person, and "anything above $8–10 per person would require the hack to be something really astonishing," says the hacker. To reduce costs, some hackers forage in trash bins and scrap piles of building materials.

Time is also a valued student commodity. Although something as simple as a banner may require more than 100 hours to make, individual hackers hope to spend no more than 10 or 20 hours on construction; most students cannot afford to spend more than a few hours per week. Consequently, larger hacks often require the involvement of many students to spread out the workload, and some hacks may take months to complete. As with any project involving a large number of people, this necessitates careful management, coordination, and planning. Luckily for the hacking process, many students tend to be willing to spend an entire night devoted to deploying a hack, probably because it is far more exciting than construction.

Conscientious hackers, like good engineers, are also concerned about hack safety. A hacker's worst nightmare is seeing a hack fall off a rooftop; aside from the fact that bystanders may be injured, it's an embarrassment. To prevent safety and structural problems, many hacks are deliberately "overbuilt." For example, one hacker says, "The slide rule (35 pounds of Plexiglas®) was held up by eight 300-pound test ropes, each individually tied." Outdoor hacks require special protection against the elements, especially the wind. Banners are often made of cloth, because paper banners have been known to rip in strong winds. Wind also increases the strain on ropes by turning anything with an exposed surface into a sail; hence, the Christmas lights hack had a volleyball net substrate, which has minimal wind resistance, and was tied down with weighted robes and duct tape. Still, unforeseen conditions can ruin a hack. An effort to change the color tiles on the Media Lab building ended prematurely when sunlight warped the painted wooden replacement tiles so much that they fell off. Luckily, the tiles were at ground level and posed no safety hazard.

Unlike most engineering projects, hackers have the added worry of trying to make sure they don't get stopped by nonhackers during the final installation. In engineering terms, this simply means that there are tighter-than-usual constraints on portability and the rapid assembly process. Although a hack may soak up more than 100 hours to build, ideally it should only take a few minutes to "deploy." Hackers are willing (although only grudgingly so) to wait hours for the perfect deployment moment (when no witnesses are present), but they aren't willing to spend hours installing a hack where they can be spotted. The more exposed the area, or the higher the traffic through the area, the more practice and planning is required.

This also means that a hack should be transportable, so that the hackers can quickly bring it out of storage to its intended destination at a moment's notice (ideal conditions tend to be short-lived). Extremely large hacks, such as the campus police cruiser and the two-story screw that graced the Great Dome, must be built modularly so that the parts can fit through doors and be transported to the deployment area. Banners or large nets with objects attached are rolled or folded up.

The need for rapid deployment also means that a hack must have only a few simple, readily accessible attachments to quickly and securely fasten it to the wall, ceiling, or rooftop. Having too many attachments is confusing, time-consuming, and tends to involve lots of ropes that get tangled easily. Anything more complicated than simple clamps, ropes, weights, and knots is avoided as well. The time-honored "KISS engineering principle" (keep it simple, stupid) aptly applies. Of course, a large hack supported at only a few places needs special attention to structural stability.

A final design factor conscientious hackers consider is removability. Good hacks should be removable without causing structural damage or, ideally, any damage at all. In fact, most hackers prefer to take down their own hacks, in part to retrieve equipment before it gets confiscated, and in part because they feel that they are most qualified to take their hacks down safely. Hence, means of attaching hacks to a surface are usually limited to nonpermanent techniques. This is generally in accordance with rapid-deployment techniques that stress the use of knots and weights instead of time-consuming welding torches or drilling equipment. One example of the importance of removability is the die hack, in which a cloth with dots (like those found on dice) was placed without official permission over an officially permitted large cubic structure hanging in Lobby 7. The hackers later helped both MIT Physical Plant and the creators of the cube take down the fabric safely and easily. Apparently, the hackers were inspired to help when they learned that the plan to remove the hack would have made the fabric even harder to remove than before. The speedy removal was especially important because the cube was to be displayed as part of an important meeting between an MIT research group and Japanese representatives.

Once the design has been finalized, construction begins. Luckily, finding and using construction equipment is rarely a problem. Although most hackers shy away from projects that would require thesis-level complexity, many of them are competent with and have access to power tools, sewing machines, electronics, and shop equipment. Other construction necessities, such as paint and duct tape, are even easier to acquire and use. One of the worst construction issues is space, since hacks tend to be large. "This wouldn't be so bad except hackers also usually don't want people to know there's a hack under construction," notes an anonymous hacker. Most large hacks end up spread out in living group hallways, empty classrooms, or even unused subbasements.

Once constructed, hackers try to test all the parts of the hack to make sure the Christmas lights actually light, to see that banners won't sag, or make sure tape will stick to the wall. As any good engineer knows, the failure to test something may mean finding flaws the hard way. For example, more than one unfortunate group has found that pulling something up the side of a building with a rope doesn't work because the concrete edge of a roof will cut through even thick rope. Also, the hackers need to test their own skills: knot-tying, climbing, and other deployment necessities. Though hackers would love to rehearse the deployment at least once, generally they cannot risk practicing at the actual site and must make do with off-site practices and verbal descriptions of what is to happen.

After design, construction, and testing, a hack is ready to be deployed. With luck, all goes as planned. An example of a fast successful deployment is the attempt to change the tile colors on the Media Lab. From the all-clear starting signal, to the placing of ladders and fake wooden tiles on the wall, to the removal of the last ladder, was a mere three minutes. Another successful deployment took longer: the Christmas lights hack took more than 20 minutes, time mostly spent making sure that a pile of boxes painted as presents were safely glued to each other. According to one hacker, the problem was that they "hadn't taken into account the fact that glues tend to dry more slowly in cold weather." Still, 20 minutes wasn't too bad because, as another hacker put it, "rooftops are relatively low-visibility areas." Of course, the Christmas lights weren't turned on until the hackers were ready to leave.

Assuming the arrival of campus police hasn't halted the deployment prematurely, the hackers do final checks on their hack, and then slip away, leaving MIT, and sometimes the rest of the world, to marvel at their handiwork. Perhaps only fellow hackers, however, can truly appreciate the effort, ingenuity, and planning that has gone into the design, construction, and implementation of the best hacks. It is engineering in action.

Eri Izawa is an MIT physics graduate (1992 SB). She has worked in fields ranging from computer game design, software engineering, web mastering, art, and writing, to playing useless computer games.

WHERE THE SUN SHINES, THERE HACK THEY
Samuel Jay Keyser, 1996

The title of Brian Leibowitz' historical compendium of MIT hacks, *The Journal of the Institute for Hacks, TomFoolery & Pranks at MIT* (MIT Museum, 1990) is itself a hack. Embedded in it are the initials IHTFP, which, as everyone at MIT knows, stand for "I hate this f*&^ing place." This is not the acronym's only "public" commemoration. The Class of 1995 changed the date embossed on the Dome image in the class ring from MCMXVI to IHTFP, something obvious only with a magnifying glass or a sharp eye. Earlier classes have done similar recodings of the MIT ring.

During my years as associate provost for institute life, many of my colleagues approached me with this question: If students hate this place, then why don't they just plain leave it? It is a good question to which, I think, there is a good answer: They DON'T hate this place. But if they don't, the conversation continues, why say they do? An equally good question.

The answer lies, I believe, in unpacking the hacking. When we do, we find the practical-joke-cum-parody lurking beneath. The practical joke is physical in character. One does not tell practical jokes. One plays them. Similarly, one does not tell hacks. They, too, are played. Here is how Arthur Koesler describes the practical joke in his *Encyclopedia Britannica* article:

The coarsest type of humour is the practical joke: pulling away the chair from under the dignitary's lowered bottom. The victim is perceived first as a person of consequence, then suddenly as an inert body subject to the laws of physics: authority is debunked by gravity, mind by matter; man is degraded to a mechanism.

The operative words here are "authority debunked." The hack is a physical joke designed to do just this. But it is not any physical joke. Hacks have a strong element of parody in them. They are physical jokes that parody the honest work of an Institute grounded in science and engineering. That is why MIT hacks, unlike hacks at other institutions, always have a strong engineering component. They make fun of engineering by impersonating it and then pulling the seat out from under. MIT hackers typically don't throw pies or wrap underwear around statues of founding fathers. Rather, they make large objects appear in inaccessible places, rewire lecture hall blackboards to go haywire when the instructor tries to use them, replace chiseled wisdom on friezes with silly sayings in what appears to be identical script and then do so so cleverly that it takes a SWAT team of trained rappellers to dismantle them.

Why does MIT hacking have such a long half-life? The answer lies in something called "disobedient dependency." In order to stay in a dependent relationship that is both desirable and yet threatening, one coping mechanism is disobedience. It distances the dependency, makes it bearable. Let me give an example drawn from my experience as a housemaster at Senior House. During the 1980s President Gray and his wife gave garden parties for the parents of incoming freshmen. The president's garden was filled with incoming sons and daughters and their parents. Several Senior House students took this as an opportunity to be ostentatiously disobedient. They would dress as grungily as possible. Then they would scale the wall separating the Senior House courtyard from the President's House garden and mingle with the well-dressed, well-scrubbed guests, scarfing crabmeat sandwiches as if they were auditioning for the part of John Belushi in a remake of *Animal House*. The more outrageous the behavior, the better. Some of the more inventive students would dress up as characters from *The Rocky Horror Picture Show*. Most, however, did not, attempting to *épater le bourgeois*, as it were, without props. More often than not, someone would dump a bottle of detergent in the garden fountain in order to intensify the nuisance value of his or her presence.

The superficial motive behind such "disobedience" was to embarrass those in authority, the president, his spouse, the various deans, and housemasters who showed up for the occasion. The crashers were declaring their independence from the Institute and all its folderol. The deeper motive was to provide distance between themselves and the Institute so that its judgments of them, upon which they deeply depended, would be less painful when they were made.

Why do I say that students deeply depend on the Institute's judgments of them? The reason is that the values of the students and of the faculty are the same. For the most part, the faculty are the best at what they do. The students come here to be like them. When the faculty grades them, those judgments can be painful because the students believe they are true. At some level our students know that while they are all in the top five percent of their high school classes, they will soon be recalibrated downward. I say "at some level" because a poll taken not too long ago asked the incoming class how many of them thought they would end up in the top quarter. Something like 75 percent said they would! At least half of those responding were about to discover they were not as good as they thought, not an easy pill to swallow at any stage of one's life.

Unlike the extreme kinds of disobedience that one often finds in living groups, the hack is a socially acceptable form of disobedience. It is easily distinguished from its more extreme counterparts by three properties. Hacks are (1) technologically sophisticated, (2) anonymous, (3) benign. They are technologically sophisticated because they need to parody an MIT education. They are anonymous because were they otherwise, the Institute might be forced, if only for safety reasons, to do something about them. They are benign because their goal is not to inflict pain, but to cope with pain inflicted. They do this by making fun of the Institute, diminishing it, bringing it down to size so that its judgments are brought down to size as well.

The hack is a pact that the Institute and its students enter into. Keep it anonymous, harmless, and fun and MIT will look the other way. It will even be mildly encouraging because it recognizes, as do the students, that students need to turn the Institute into an adversary. This, by the way, is why the adversarial undercurrent between students and the Institute won't go away, no matter how supportive student services are or how solicitous our staff might be or how accessible the faculty makes itself.

The hack isn't the only buffering mechanism. Another is the special relationship that students have to their living groups. Why does where a student lives take on such monumental proportions at MIT? Part of the answer is that living groups function much like disobedience; namely, as a kind of protection against the slings and arrows of institutional judgment. Living groups are safe houses, ports in a storm, raingear to keep them dry once the firehose is turned on. This them/us division is so profound, in fact, that long after they have graduated, students talk in terms not of having been at MIT but rather of having been at Senior House, or Sigma Chi, or MacGregor. MIT tacitly acknowledges this as well, which is why changing the very peculiar system of residence selection called R/O is like pulling teeth. The buffering function of the resident system is as much a part of an MIT education as are the General Institute Requirements. The same is true of hacks.

Hacks and living groups, then, are to the Institute what sunglasses are to the sun: a form of protection that makes it possible to live with the light. Not every student hacks. Not every student feels the same degree of disobedient dependency. But every time hackers help to place a police car on the dome, they are providing shade in a very sunny clime.

POSTSCRIPT, 2010

In 1996, when this article was written, the hack of the year was a beanie installed on the Great Dome, with a propeller that spun in the wind. Fourteen years later (2010) the torii gate outside the Media Laboratory was adorned by an upside-down room hanging from the underside of the arch, complete with an unfinished game of pool, a whiskey bottle on a side table, and a sleeping cat. Obviously, the tradition of imaginative, intricate, funny hacks remains alive and well at MIT.

While the hacking tradition lives, another tradition has died. This one dates back to the late 1960s and persisted right up through the 1980s. I am thinking of the tradition of student protest, the tradition that roiled the MIT campus from the war in Vietnam through to the antiapartheid divestment protests. It can't be for lack of issues that the protests have died. Think of shock and awe in Iraq, to name just one. Protests are as unthinkable now as would be a return to showing pornographic movies on registration day.

While the culture outside of MIT has changed radically, the culture within MIT, the one that leads students to cope with the rigors of an MIT education by ridiculing it, hasn't changed one bit.

Samuel Jay Keyser is professor emeritus of linguistics and holder of the Peter de Florez chair emeritus at MIT. He was associate provost for institute life from 1986 through 1994 and came by his experience with hacking during those halcyon years and during his tenure as housemaster of Senior House. He returned to his faculty duties in 1994 and retired, exhausted, four years later, though he still serves as special assistant to the chancellor.

GLOSSARY OF MIT VERNACULAR

10–1000 Unofficial room number of the Great Dome

1893 dormitory Now known as the East Campus dormitory

APO Alpha Phi Omega, a service fraternity

ATO Alpha Tau Omega fraternity

Athena Central MIT computer network

Baker House Dormitory

Beast from the East Second Floor East in the East Campus dormitory

Beaver MIT's mascot

Brass Rat MIT's school ring ("rat" refers to the beaver on the ring)

Bruno Unit of measurement resulting from the free fall of a piano (see The Numbers Game)

Burton House Dormitory

Class of '93 dormitory Now known as the East Campus dorm

Coffeehouse Club Informal affiliation of hackers who meet regularly to explore the Institute and share hacking knowledge

Cruft Obsolete junk that builds up in a home or office

DKE or Deke Delta Kappa Epsilon fraternity

Dome There are two, "The Great Dome" over Building 10 and a smaller dome over Building 7 (see Domework)

East Campus Dormitory (originally called the '93 dormitory or Class of 1893 dormitory)

Fishbowl Originally located off the Infinite Corridor where MIT Student Services offices are now located. It was the most visible of the Athena public-computing clusters.

Great Dome Larger of the two Institute domes, over Building 10

Greenspeak Text displayed on the windows of the Green Building by strategically turning lights on and off and raising and lowering blinds in the outward-facing offices

Hack An inventive, anonymous prank (see Hack, Hacker, Hacking)

Harvard Bridge Bridge connecting Boston to Cambridge at MIT

Hell MIT

Hosed Overwhelmed with homework

IAP Independent Activities Period, in January between the fall and spring terms

I.H.T.F.P. "I Hate This F*&^ing Place" and infinite variations (see Intriguing Hacks to Fascinate People)

IHTFP Gallery Online gallery of MIT hacks at <http://hacks.mit.edu>

ILG Independent living group

Infinite Corridor Quarter-mile-long corridor running through several connected buildings that cuts through the heart of the Institute

Jack Florey or Jack E. Florey Hacking group from the East Campus dormitory (Fifth East)

James E. Tetazoo Hacking group from the East Campus dormitory (Third East)

Killian Court Large courtyard facing the Great Dome on one side and the Charles River on the other (graduation ceremonies take place here)

Larry Archenemy of James Tetazoo and champion of Elvis, this hacking group congregates around the forty-first floor of the western front of East Campus

Lobdell Cafeteria in the student center

MacGregor House Dormitory

ORK Order of Random Knights, a hacking group from the Random Hall dormitory

R/O Residence and Orientation Week for incoming students

Rush Period during R/O when new members were recruited for living groups (discontinued in 2001)

Senior House Dormitory

Smoot Unit of length used to measure the Harvard Bridge (see The Numbers Game)

Squanch Cartoon character representing Third East in the East Campus dormitory

Tech MIT (outdated)

The Tech Student newspaper founded in 1881

Tech Talk Campus newspaper published by the MIT News Office from 1957 to 2009

THA Technology Hackers Association, a now-defunct hacking group and one of the preeminent forces in the annals of hacking

Tool To study, or one who studies too much

SOURCES

Much of the material for this book derives from the Hack Collection that is part of the General Collections of the MIT Museum. This rich repository of photographs, documents, and artifacts is available to interested researchers by appointment. See the website for more information: <http://web.mit.edu/museum/collections/research.html>.

The student newspaper, *The Tech*, has digitized its entire collection of newspapers dating back to 1881. This is the indispensable record of student life, and researchers can easily search its archives for hacks and other pranks: <http://tech.mit.edu/browse.html>. MIT's student-produced handbook, *HowToGAMIT* (How to Get Around MIT), contains all the information that students really want to know about surviving and thriving at MIT. There is a lengthy section on hacking that gives a bit of history as well as tips, hacker ethics, and other vital information for students thinking about hacking.

The best source of information on recent hacks is the IHTFP Gallery website: <http://hacks.mit.edu>. The MIT Museum gratefully acknowledges its debt to the curators and contributors of this wonderful digital collection of hack history. For detailed information about the 2006 Caltech Cannon hack, please see <http://www.mitcannon.com> for numerous links to the media reports cited in Eric Bender's essay. Additional material may be found at the IHTFP Gallery website and the online archives of Technology Review: <http://www.technologyreview.com>.

This is the fourth book the MIT Museum has published on the history of hacking at MIT. One of the special features of each volume has been the wonderful essays contributed by members of the MIT community. The listing below indicates the publication in which each essay first appeared.

1. Leibowitz, Brian M. *The Journal of the Institute for Hacks, TomFoolery & Pranks at MIT.* Cambridge: The MIT Museum, 1990.

"Hack, Hacker, Hacking" excerpts from the book's Introduction, pp. 2–3; "The Great Breast of Knowledge," p. 14; "USS *Tetazoo*," p. 87; "No Knife" by James Tetazoo, p. 85; "Student Leaders at MIT Claim Harvard as Colony" by Jessica Marshall, p. 97. Reprinted from Harvard *Crimson*; and "Hacking by Mail" excerpts from *Building Cozy*, pp. 138–140.

2. Haverson, Ira, and Fulton-Pearson, Tiffany, editors. *"Is This The Way To Baker House?"*: *A Compendium of MIT Hacking Lore.* Cambridge: The MIT Museum, 1996.

"Father Tool's Grand Tour," pp. 32–36; "Green Eggs and Hair," pp. 73–74; "Why Ruin the Atrium," p. 72; "The Case of the Disappearing President's Office" by Charles M. Vest, pp. 128-129; "Engineering in Action" originally titled "Producing a Hack" by Eri Izawa, pp. 86–90; "Where the Sun Shines, There Hack They" by Samuel Jay Keyser, pp. 102–105.

3. Peterson, T. F. *Nightwork: A History of Hacks and Pranks at MIT.* Cambridge: MIT Press in association with the MIT Museum, 2003.

"Why We Hack," anonymous, pp. 169–170; "Recalculating the Infinite Corridor" reprinted from IHTFP Gallery, <http://hacks.mit.edu/Hacks/by_year/1997/infinity_rods>; "It's Not a Job, It's an Adventure," by David Barber, pp. 155–158 based on an earlier essay titled "The Fine Art of Hack Removal" by David Barber, p. 119 in *"Is This the Way to Baker House?"*; "Mastery over the Physical World: Demonstrative and Pedagogical Value of the MIT Hack," by André DeHon, pp. 160–161.

One final note: Hacks are more-or-less clandestine, so the accounts and quality of the photographs vary considerably. Every attempt has been made to verify the information in this book, but sorting out fact from embellishment is sometimes difficult. Over the years, many materials in the MIT Museum collection have been donated anonymously. We have done our best to give credit to the right source but will accept happily any corrections and emendations for future editions.

PHOTO CREDITS

x–xi map by Margaret Mazz; **2** Chuck Kane; **9** Rob Radez; **10** MIT Museum General Collections; **14–17** Erik Nygren; **18–19** Eric Schmiedl; **20** Grant Jordan; **21** Eric Schmiedl; **22** TungShen Chew; **23** Greg Perkins; **25** Alessondra Springmann; **26** Erik Nygren; **28** Ray C. He; **29** Robert Gens; **30** Eric Schmiedl; **31** Eric Schmiedl; **33** Deborah Douglas; **35, 37** Eric Schmiedl; **39** MIT Museum General Collections; **40** Joe Dennehey, *Boston Globe*/Landov; **41** Permission to reprint from the *Technique*: Ken Haggerty, managing editor 2010–2011 / Photo by Warren Kay Vantine Studio, Boston; **43** (both images) Simson L. Garfinkel; **44** MIT Museum General Collections; **45–46** Frank Schoettler, negatives provided by Debbie Levey; **47–49** (all images) Brian Leibowitz; **51** (top) Donna Coveney/MIT News Office; **51** (bottom) MIT Museum General Collections; **52** Bob Schildkraut VI, Class of 1962; **53** Donna Coveney/MIT News Office; **54** Marc PoKempner/MIT News Office; **55** Jon Williams; **57** MIT Museum General Collections; **58** Carl Bazil; **59, 62–63** (all images) MIT Museum General Collections; **64–65** Alan Devoe/*Technique*; **66–67** Donna Coveney/MIT News Office; **68** (top) Douglas Keller; **68** (bottom), **69–71** (all images) Terri Iuzzolino Matsakis; **72** William Litant; **73** Erik Nygren; **75** Terri Iuzzolino Matsakis; **77** Calvin Campbell/MIT News Office; **79** Frank Hill/*Boston Herald*; **80–81** William C. Hoffman and Theta Chi Fraternity; **81** (right) Chun Lim Lau; **82** (both images) Tien Nguyen; **83** Mark Virtue; **85** Simson Garfinkel/MIT News Office; **86** Donald M. Davidoff ('86); **87, 89** Terri Iuzzolino Matsakis; **90** (all images) Eric Nygren; **92** (top) Brian Leibowitz; **92** (bottom) Eric Nygren; **93** André DeHon; **94–95** (both images) Terri Iuzzolino Matsakis; **96** André DeHon; **98–99** (all images) © Michael J. Bauer/Eri Izawa, all rights reserved; **101** (both images) Terri Iuzzolino Matsakis; **103** Simson L. Garfinkel/*The Tech*; **104** Terri Iuzzolino Matsakis; **105–106** Brian Leibowitz; **108** (top) Marc Horowitz; **108** (bottom) Donna Coveney/MIT News Office; **112** MIT Museum General Collections, the Escher inset on right image is courtesy of M. C. Escher's "Relativity" © 2010 The M. C. Escher Company—Holland, All Rights Reserved, <http://www.mcescher.com>; **114** (top) Brian Leibowitz; **114** (bottom) MIT Museum General Collections; **115, 117** (both images) Terri Iuzzolino Matsakis; **118** Ron M. Hoffmann; **119** Chris Laas; **122–124** (all images) Terri Iuzzolino Matsakis; **125** Donna Coveney/MIT News Office; **126** MIT Museum General Collections; **127** Terri Iuzzolino Matsakis; **129** David M. Watson; **130** Jeff Thiemann; **131** Larry-Stuart Deutsch, MD, MIT Class of 1967; **132** Carl Lacombe; **133** Peter Büttner; **134** MIT General Collections; **135** Steve Slesinger, Class of 1975; **136** Brian Leibowitz; **137** Oren Levine; **138–139** Brian Leibowitz; **141** J. Alan Ritter; **143** John E. Ohlson; **144** Brian Leibowitz; **145** (left) Brian Leibowitz; **145** (right) MIT Alumni Association and the MIT Club of Boston; **147** Brian Leibowitz; **148, 151** (top) Terri Iuzzolino Matsakis; **151** (bottom) Erik Nygren; **153** Donna Coveney/MIT News Office; **155, 157** MIT Museum General Collections; **159** Brian Leibowitz; **163** C. E. Pence; **167** Heather Howard Chesnais; **169** Mark James; **171** Donna Coveney/ MIT News Office; **173** (both images) *Boston Globe*/Landov; **174–175** (all images) Bob Brooks; **177** Mike Bernard; **179** MIT Museum General Collections; **180** Charles A. Honigsberg; **181** Dean Phillips; **182** (left) Dean Phillips; **182** (right) Gene Dixon, courtesy of the *Boston Herald*; **184–186** (all images) Eric Nygren; **188–189** (both images) MIT Museum General Collections; **191** Al Paone; **194, 211** Terri Iuzzolino Matsakis

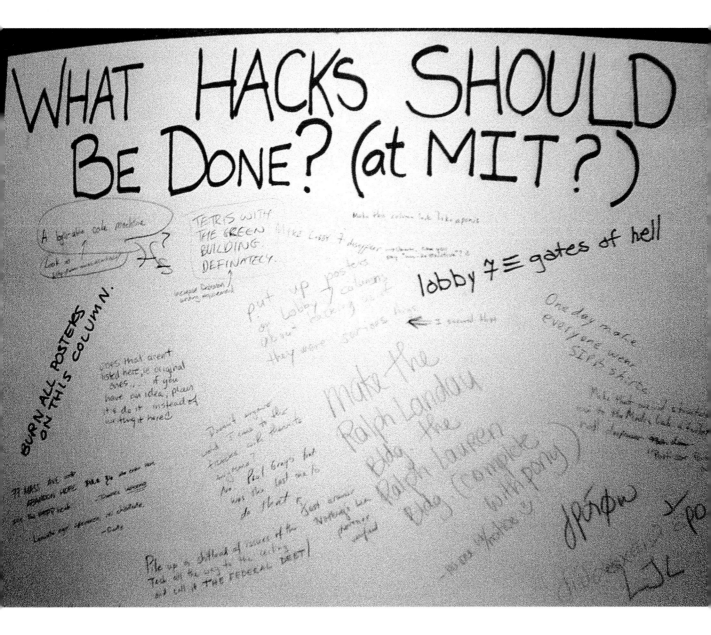

INDEX OF HACKS